A PHOTC

B[
O[THE
CAYMAN ISLANDS

A PHOTOGRAPHIC GUIDE TO THE

BIRDS
OF THE
CAYMAN ISLANDS

PATRICIA E. BRADLEY

PHOTOGRAPHY BY YVES-JACQUES REY-MILLET

CHRISTOPHER HELM
LONDON

For Theodora and Raphaela

For Alexandra

Published in 2013 by Christopher Helm, an imprint of Bloomsbury Publishing Plc.,
50 Bedford Square, London WC1B 3DP

www.bloomsbury.com

Bloomsbury Publishing: London, New Delhi, New York and Sydney

A CIP catalogue record for this book is available from the British Library

Commissioning Editor: Nigel Redman
Project Editor: Jim Martin
Design by Julie Dando, Fluke Art

ISBN (paperback) 978-1-4081-2364-5
ISBN (hardback) 978-1-4081-9399-0
ISBN (epdf) 978-1-4081-8903-0

Printed in China by C&C Offset Printing Co., Ltd.

This book is produced using paper that is made from wood grown in managed sustainable forests.
It is natural, renewable and recyclable. The logging and manufacturing processes conform to the environmental
regulation of the country of origin.

10 9 8 7 6 5 4 3 2 1

Front cover (clockwise, from top left): Vitelline Warbler, Cuban Parrot, Cuban Bullfinch, West Indian Woodpecker, Red-footed Booby. All ©Yves-Jacques Rey-Millet.

CONTENTS

Bananaquits feeding.

PREFACE

This is the first photographic guide for the Cayman Islands that describes and illustrates all regular species of breeding and migrant birds. It is hoped that it will bring a greater understanding of the avian biogeography of these unique islands, as well as illuminating some of the local mysteries of migration and endemism. Like all such endeavours this book is built on information gathered between the first checklist in 1882 and the last in 2000, as well as on new and changing status data provided by the authors and the Bird Club.

Despite their small size and low elevation, the Cayman Islands attract a disproportionate number of species due to their geographical position between North America and South America. Breeding species number 50, three of which are summer-breeding migrants, the remainder being residents. A further 195 species are non-breeding migrants. Species new to the islands will continue to be identified as North American migrants, en route to Middle America (Mexico and Central America) and South America via the Gulf of Mexico, are deflected to western Cuba and the Cayman Islands in the western Caribbean Sea. Some birds overwinter, others stop-over for several days or weeks to replenish fat reserves, while others are only observed as they overfly the islands. All migrants are journeying northwards in spring to their breeding grounds or southwards in fall to their wintering grounds.

Local knowledge helps, so read the Introduction! Identifying the different habitats described allows species to be pinpointed in preferred sites. In many cases the differences are subtle: this is a land with no high mountains or flowing rivers, where a species will make a distinction between feeding on a freshwater rain pool but not on a saline pond. There is also a distinction between dry forest and open-canopied shrubland when it comes to locating breeding endemic landbirds. The Near-threatened Cuban Parrot – which we claim as the Cayman Parrot on Grand Cayman and the Brac Parrot on Cayman Brac – requires forest to breed, only nesting in holes in trees of a certain diameter. Being able to identify fruiting trees makes for an easy day's birding as both insectivorous and fruit-eating birds, breeding or migrant, flock to the ripe fruits of Pepper Cinnamon and Wild Fig.

If possible, visit the three Islands: Grand Cayman, Little Cayman and Cayman Brac. The flight on a small plane makes it easy to appreciate the differences between each island as well as enjoying a 'birds-eye view'. The sister islands of Little Cayman and Cayman Brac have wonderful seabird colonies, their own endemics, good swimming, local crafts and fine food. Perfect places to chill. Good birding!

Patricia E. Bradley and Yves-Jacques Rey-Millet
May 2012

Male Scarlet Tanager: Regular passage migrants to the Cayman Islands.

ACKNOWLEDGEMENTS

Our special thanks to Arturo Kirkconnell for insightful comments in his review of the text and images, a huge task for which we are most grateful. We thank also to John Bothwell for his valuable comments on the Introduction, Gina Ebanks-Petrie for her expertise on the Conservation section, Stuart Mailer for improving the Where to Watch Birds section and Paul Watler for providing the latest nomenclature for Appendix 2. Jeremy Olynik spent many hours creating the maps as well as providing all topography and land areas – my deep appreciation to him and thanks to the Department of the Environment for permission to use them. All their contributions have improved the text immeasurably and we are most appreciative of their support. Any remaining errors are ours.

Our thanks to all the Bird Club members who have generously shared field notes, and especially Peter Davey and Trevor Baxter. Thanks also to Stuart Mailer, Frederic Burton and Nancy Norman for sharing photographs. We thank Larry Manfredi in Florida for his help and Wallace Platts on Cayman Brac for support, and Cayman Helicopters for enabling us to shoot habitat images. We are grateful to landowners for allowing us access to the best photographic sites on Grand Cayman: John Lawrus, Director of the Botanic Park, Albert Estanovich, Handal Whittaker, William Peguero, and Anton and Geraldine Duckworth.

Special thanks to Alexandra Guenther-Calhoun for time in the field, her excellent sonograms of endemics and for the many hours filing around 50, 000 bird images into species groups and individual folders, and making the final assembly of the book so enjoyable. And finally we celebrate 30 years of working together: a third field guide and the Checklist.

We are grateful to Julie Dando for a superb job in interweaving the photographs and text, and we thank Nigel Redman and Jim Martin for bringing the book to a successful conclusion.

Patricia E. Bradley and Yves-Jacques Rey-Millet
Cayman Islands, July 2012

Patricia E. Bradley and Yves-Jacques Rey-Millet in the field.

INTRODUCTION TO THE CAYMAN ISLANDS

HISTORY

> 'We sighted two very small and low lying islands, full of tortoises, as was all the sea about, insomuch that they looked like little rocks, for which reason these islands were called Tortuous.'

From the narrative on the fourth and last voyage of Christopher Columbus who, on 10 May 1503, claimed the Cayman Islands for Spain. They were later named *'Los Caymanas'* after the Carib word for crocodile. No archaeological evidence of permanent settlement by the pre-Columbian Taino people of the Greater Antilles has ever been uncovered. The Cayman Islands continued to remain uninhabited for a further 240 years because they lacked mineral wealth, such as gold, and were not a source of slaves. However, due to their position close to the sea route between the New and the Old Worlds, the islands became an important provisioning stop for Spanish ships to take on fresh water, live turtles, iguanas, seabirds' eggs, and wood. These vessels were trailed by ships of privateers and buccaneers from England, France and The Netherlands, including Sir Francis Drake and the infamous 'Blackbeard'. The first settler families arrived on Grand Cayman from England via Jamaica in the 1740s and 100 years later the population had reached 2,000. Little Cayman and Cayman Brac were first settled c. 1833.

GEOGRAPHICAL POSITION

The Cayman Islands comprise three islands and a few small cays situated close to the Greater Antilles in the extreme western Caribbean Sea between latitudes 19° 20′N and 19° 43′N and longitudes 79° 50′W and 81° 21′W. They lie c. 700km (435 miles) south of Miami, Florida, 280km (174 miles) northwest of Jamaica and 240km (150m) south of Cuba. The Yucatan Peninsula, Mexico is the closest landfall on continental America. Grand Cayman lies c. 130km (80miles) south-west of Little Cayman and Cayman Brac, which are separated from each other by a 7km (5 miles) wide channel.

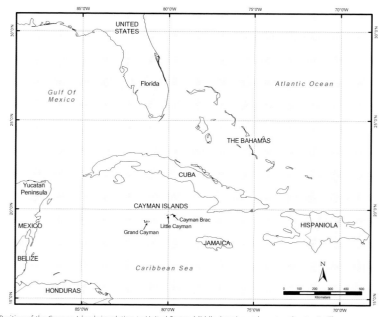

Position of the Cayman Islands in relation to United States, Middle America and western Greater Antilles.

AREA, ELEVATION AND POPULATION

Grand Cayman

An area of 197km ^{2} (99 miles^2) and a population of *c.* 53,000 in 2011, it is the largest and most westerly of the Islands. It is 35km (25 miles) long and *c.* 8km (5 miles) wide with a maximum elevation of 19m (62ft). It supports 96% of the total population of the Cayman Islands. The capital is George Town, with population centres at West Bay, Bodden Town, East End and North Side; the majority of the population lives on the western half of the island.

Little Cayman

It has an area of 28km^2 (10miles^2) and a population of about 200 in 2011. It lies 7km (5m) to the west of Cayman Brac and 117km (73miles) to the east north-east of Grand Cayman. It is 16.3km (6miles) long and 1–3km (0.6–1.2miles) wide with a maximum elevation of 14m (43ft). Blossom Village is the main settlement. Reefs encircle the island, except at the western end where a natural deep water port is situated.

Cayman Brac

It has an area 38km^2 (13miles^2), a population of 1,900 in 2011, and it is the most easterly of the Islands. It is 20km (8miles) long and 1-3km (0.6-1.2miles) wide, with a maximum elevation of 47m (153ft), the highest point in the Cayman Islands. The population is centred in the west and along the northern coastal shelf at West End, Cotton Tree Bay, Stake Bay, Watering Place, Creek and Spot Bay.

Government

The Cayman Islands are an Overseas Territory of the United Kingdom.

Language

English. Some speak an English-West Indian patois, and there are sizable Filipino and Spanish-speaking groups.

CLIMATE

The Islands lie in the central Caribbean basin, which is demarcated on the north and east by the Greater and Lesser Antilles. These offer some protection from the north-east trade winds, which persist for most of the year, although some south-easterlies blow in summer. There are two distinct climatic seasons: a humid wet season from May to November during which most of the annual rain falls and the mangrove and some dry forests become inundated, and a dry to semi-arid season from December to April. Cold fronts from North America bring northerly winter storms from December to March.

Rainfall is distributed unevenly due to the islands' topographies, elevations being higher in the west than in the east; the average annual rainfall is 1,351mm on Grand Cayman, 1,174mm on Little Cayman and 1,111mm on Cayman Brac. However, drought conditions do occasionally occur. Average daily air temperatures on Grand Cayman range from 20–28°C in January to 25–33°C in July. Sea temperatures follow a similar pattern. Relative humidity ranges from 65–100% and rarely falls below 40%. Similar conditions occur on Little Cayman and Cayman Brac.

The Islands lie within the catastrophic hurricane zone and these have occurred mainly between August and November since the beginning of the 20th century, usually affecting either Grand Cayman or Cayman Brac and Little Cayman, but seldom all three together. High winds cause catastrophic felling of forest and drive sea surges inland resulting in inundation and rapid die-off of the mangrove vegetation, as has occurred at Tarpon Lake, Little Cayman. Hurricanes significantly alter the topography by damaging frame-building corals on the reefs and by depositing boulders in a series of ramparts along some coasts. Major devastation was caused to Grand Cayman in 2004, Cayman Brac in 2008 and Little Cayman in 2005 and 2009.

The Islands lie in the path of the Caribbean Current that flows north-west at speeds of 0.5m per second to the Strait of Yucatan and the southern Gulf of Mexico. Ocean currents are known to transport vegetation rafts and seeds from Cuba and Jamaica to Cayman Brac or Little Cayman in about nine days, and the currents and eddies around Little Cayman deposit debris from Cuba on both its north and south coasts. Rafting from these islands probably accounts for colonisation by many species of flora and fauna.

Grand Cayman, north coast showing the bluff cliffs.

GEOLOGY

The appearance of the surface rocks in the Cayman Islands, where rock is either jagged and black or forms large blocks as on the Cayman Brac bluff, leads many to believe their origin must be volcanic. This is not the case. The Islands are comprised of limestones derived from the sea floor.

The three Islands are emergent peaks situated on the submarine Cayman Ridge, which defines the southern boundary of the North American Plate, stretching from the Sierra Maestra of Cuba to the Gulf of Honduras. The Cayman Ridge forms the northern boundary of the Cayman Trench that reaches depths in excess of 6,000m; to the north of the Ridge is the 4.500m-deep Yucatan Basin. The Mid Cayman Rise, which is an active spreading centre tectonically active for at least six million years, is located to the south-west of Grand Cayman at 82°W where the Oriente Transform Fault delineates the boundary between the North American Plate and the Caribbean Plate.

It seems likely that the Cayman Islands were isolated from the Pacific by the end of the Pliocene since the older strata of the islands from the Oligocene, Miocene and late Pliocene (from 30 to 3 million years old) contain coral fauna of the Pacific realm, while the younger strata have a typical Caribbean coral fauna. Each Island lies on a separate fault block and is surrounded by ocean deeps in excess of 2,000m and none were ever connected to any landmass. The central part of each island is formed of an exposed platform of hard limestone rocks, named Bluff dolostone, which has been influenced strongly by sea-level and climatic changes, rather than by tectonic changes. Parts of each Island are thought to have been above sea level for at least two million years. The surface of the Bluff rock is deeply jointed, highly porous and typically has a jagged phytokarst surface with honeycombed pinnacles, fissures and sinkholes that makes walking in the interior of each Island hazardous. The root systems of the dense vegetation penetrates the rock and also channels rainwater that dissolves the limestone, forming extensive cave systems and deep cavities, a rich source of avian subfossils: many of them from the prey of owls.

The ancient limestone core of each Island is overlain by Ironshore Formation exposed as a coastal terrace 4m (13ft) above mean sea level and occurring over much of the western half of Grand Cayman and Little Cayman and around all the coasts of the three Islands. It dates from the late Pleistocene (124,000±8,000 years ago) and was deposited during the last interglacial period when sea level was c. 6m (20ft) above present levels. The large caves above the wave-cut notch on the north and south coasts of Cayman Brac were formed at this time. The Ironshore Formation is dark-coloured with a hard calcrete crust over the underlying softer, pale, oolitic limestones, coral and molluscan fossils and lagoonal muds. On Grand Cayman, the Ironshore was laid down in an enlarged lagoon bounded by

The Ironshore, from the Pleistocene, forms a coastal terrace around the coastline of the three Islands. West coast, Grand Cayman.

patch reefs and its thickness varies from 9–17m (27–50ft); the present North Sound represents the remnant of this lagoon. Mangrove swamps began to form during the Holocene (from *c.* 9,000 years ago) and the mangroves became extensive in western Grand Cayman and Little Cayman as sea levels rose *c.* 4,500 years ago.

ORIGINS OF THE CAYMAN ISLANDS AVIFAUNA

In the Greater Antilles mammal fossils from the Eocene (65 million years ago) suggest the presence of an early land-bridge from Jamaica to Central America, before Jamaica became submerged between the mid-Eocene until the Miocene. Between the Pliocene (3–5 million years ago) to the Present, when the West Indian islands were substantially in their present shape and positions, later colonisers arrived by over-water dispersal, which explains the depauperate and unbalanced nature of the avifauna in relation to the mainland. Jamaica may have been the first island colonised by the majority of new taxa as it was closest to the mainland, with exposed banks and sea mounts greatly reducing over-water distances in periods of lower sea level. Also the Cayman Islands and the Swan Islands were both closer to the mainland and it is likely that birds reached western Cuba via these islands, in addition to the route from the Yucatan. Parts of the Cayman Islands have been above water for 2–3 million years.

During the interglacial periods in the Pleistocene *c.* one million years ago, when the Grand Bahama Bank became almost contiguous with Cuba, taxa entered the West Indies from temperate North America via Florida, the Bahamas and Cuba: e.g. Northern Mockingbird and Northern Flicker. A few South American species moved northwards into the Lesser Antilles and, although few have penetrated north or west of Guadeloupe, the Caribbean Elaenia may have colonised from the east. Recent studies on the systematics of West Indian endemics show them to be often very diverse from their mainland relatives. Also, where a species is found on more than one island, the individual populations are also genetically diverse from each other, suggesting either that the West Indian lineages are relatively old and long isolated or that they have evolved more rapidly than than their counterparts on the mainland.

In the late Pleistocene glaciations (c. 17,000 years ago), the elevations and surface areas of the Cayman Islands were greater, as sea levels had dropped to c. 120m (380ft) below today's mean sea level and xeric (dry) habitats had expanded considerably throughout the northern Caribbean. Around 10,000–12,000 years ago, the Pleistocene-Holocene transition began, causing a rise in temperatures, rainfall and sea-levels and resulting in the climate changing from dry to wet (mesic). On Grand Cayman and Cayman Brac, several species became extinct: of 34 species identified from avian fossils (14,000–4,500 years ago), four are now extinct (two giant raptors, a quail-dove and a new species of bullfinch from Cayman Brac), 22 are still present and eight species occur elsewhere although they are no longer present in the Cayman Islands (Morgan 1994). This confirms the close relationship of the earlier Cayman avifauna with that of Greater Antilles and the Bahamas.

The Pleistocene avifaunas of Grand Cayman and Cayman Brac were also much more homogeneous, when the Northern Flicker, Western Spindalis, Cuban Bullfinch and Jamaican Oriole were present on both Islands, whereas today none occur on Cayman Brac.

Human colonisation of previously uninhabited islands often leads to rapid decline, extirpation and extinction of their faunal populations. Morgan (1994) estimated that at least five now extirpated species survived into the post-Columbian era, including large colonies of Audubon's Shearwater: identified from well-preserved feathers and subfossils on Cayman Brac. There is no historical record of its occurrence and it was probably extirpated early in the 19th century before permanent human settlement, partly by visiting seafarers and partly by predation by rats and cats that had escaped from their ships.

Island biogeography theory proposes that species composition of insular faunas is due to a balance ('turnover') between colonisation by long-distance dispersive species and the extinctions of others, resulting in a form of equilibrium where the number of resident species depends primarily on island size and degree of isolation. It is interesting that seven new breeding birds became established in the Cayman Islands in the 20th century and post-2000: Cattle Egret in 1957, Least Bittern in mid-1990s, Bridled Tern in 1995, Yellow-crowned Parrot in 1993, Short-eared Owl in 1997 and House Sparrow in 2007. The White-winged Dove is also thought to have colonised in the 20th century as there are no records before 1935. One endemic landbird, the Grand Cayman Thrush, became extinct post-1938 and one endemic subspecies, the Jamaican Oriole, became extirpated (lost to the Cayman Islands but still present on Jamaica) (Appendix 1). This theory does not take into account the effect of human disturbance or that the populations of two landbirds were likely already vulnerable. The extinction of the thrush may have been a gradual process that began during the Holocene and it moved towards scarcity and decline following the loss of primary forest in the 18th and early 19th centuries.

VEGETATION AND HABITATS

The vegetation of the Cayman Islands is typical of small, low elevation tropical islands without mountains or rivers. The vegetation is classified into semi-deciduous dryland communities of Dry Forest (including Dry Woodland) and Dry Shrubland and wet evergreen communities of Mangrove Forest, Mangrove Shrubland and Herbaceous Wetlands (Burton 2006). Avian habitats associated with these communities are discussed below as well as habitats relating to coastal areas and areas altered by man. (See Appendix 2 for the scientific names of the dominant plants in these communities)

Dry Forest and Dry Shrubland dominate the terrestrial interior of the Cayman Islands, eking out a precarious existence on the bluff limestone karst. These communities provide the majority of terrestrial habitats for endemic, resident and migratory landbirds.

Dry Forest

This is defined as having 60–100% canopy cover. The vegetation is dry-deciduous and canopy height ranges from 5–15m (16–46ft) with a lower understorey. The once extensive forest cover of the bluff limestone on Grand Cayman was reduced by 2012 to 15% of the land area, confined mainly to the Mastic area east of the Botanic Park and scattered areas in the eastern districts. On Cayman Brac, Dry Forest until recently extended unbroken over the central bluff. In 2012 48% remained but was undergoing rapid fragmentation due to increasing urban development. On Little Cayman, up to 2010 c.75% of the forest remained pristine, covering 27% of the land area but by 2012 large areas of the eastern forest and shrubland had been sold for development.

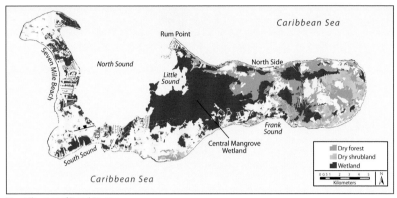

Vegetation map of Grand Cayman.

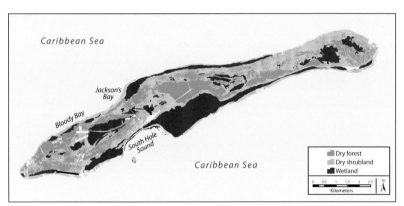

Vegetation map of Little Cayman.

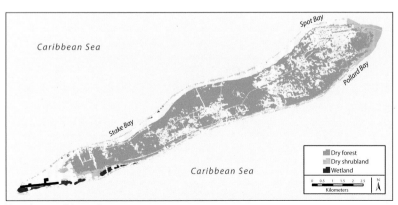

Vegetation map of Cayman Brac.

Dry Forest (right) border with Black Mangrove Forest of the Central Mangrove Wetland (left), Grand Cayman; both provide prime parrot breeding habitat.

The dominant forest trees of this community provide breeding and foraging habitat for birds. Dominant trees of the Dry Forest common to the three Islands include Red Birch, Pepper Cinnamon, Silver Thatch, Cabbage Tree, Narrow-leaf Ironwood, Ironwood, Wild Fig (two species), Bastard Ironwood, Wild Jasmine, Cherry, Smoke Wood, Manchineel, Bastard Strawberry, Yellow Mastic, Wild Sapodilla, Candlewood, Pompero, Bitter Plum, Spanish Elm, Whitewood and Bastard Mahogany. Some trees are confined to one of two islands. Royal Palm, Shake Hand (two species), Bastard Fustic and Black Mastic are only on Grand Cayman. Balsam, Fiddlewood and Cedar occur only on Grand Cayman

Royal Palm Forest, favoured parrot nesting habitat, occurs along the transition between Dry Forest and Mangrove wetlands, Grand Cayman.

and Cayman Brac and Bull Thatch Palm on Grand Cayman and Little Cayman. Wild Ginep is confined to Cayman Brac and Poison Tree to Little Cayman. Seasonally flooded Dry Forest occurs at sea level on Grand Cayman, where Royal Palm and Mahogany are dominant, and on Little Cayman, where Mahogany is dominant. Several trees (Red Birch, Wild Fig, Sea-grape, Pepper Cinnamon, Silver Thatch and Buttonwood) provide staple food for birds as they bear fruits throughout the year on each island. Bromeliads and the endemic Banana Orchid found here and in Dry Shrubland are important as nest sites and for nest-building materials for landbirds. Introduced fruiting trees established in the wild, for example, Mango, Neesberry and Tamarind, attract many landbirds, including normally insectivorous species.

Dry Woodland

This has 50% canopy cover and the vegetation is characteristic of old second growth. It is present on 'dry cays', elevated areas of limestone that support terrestrial vegetation within mangrove wetlands.

Species associated with Dry Forest habitat Much of the remaining forest in central and eastern Grand Cayman remains pristine, without roads, and is extremely rich in landbirds. It is the preferred breeding habitat for the endemic subspecies of the Caribbean Dove, Grand Cayman and Cayman Brac Parrots, West Indian Woodpecker, Northern Flicker, Yucatan Vireo, Western Spindalis, Red-legged Thrush, Caribbean Elaenia and Loggerhead Kingbird. It is also the preferred breeding habitat for other indigenous birds on all three Islands: Yellow-crowned Night-Heron (also nests in wetlands), White-crowned Pigeon,

Zenaida Dove (the White-winged Dove has made some inroads into the forest but is mainly found in urban and littoral areas), Barn Owl, Mangrove Cuckoo, La Sagra's Flycatcher and Cuban Bullfinch. Breeding in this habitat on Cayman Brac and Little Cayman are Tricolored Heron, Cattle Egret and summer migrant Black-whiskered Vireo, while on Little Cayman are Greater Antillean Grackle, Red-footed Booby, Magnificent Frigatebird and Snowy Egret. Non-breeding migrants include thrushes, Gray Catbird, Cedar Waxwing, flycatchers, vireos, over 20 species of warbler, Summer and Scarlet Tanagers, and many vagrants.

Seasonally flooded forest with Royal Palms and Mahogany trees is important for breeding parrots on Grand Cayman. Cedar in the bluff forest on Cayman Brac provides optimum parrot nest sites and good Wild Ginep flowering years are associated with successful breeding by parrots. Two plants important as food for Cuban Bullfinch are the vines Wire Wiss and *Chromolaena odorata*.

Dry Forest on the bluff at the Brac Parrot Reserve, Bight Road footpath, Cayman Brac.

Dry Shrubland dominated by Agave and Silver Thatch Palms, eastern bluff, Cayman Brac.

Dry Shrubland

This covered 7% of the land area of Grand Cayman, 37% of Little Cayman and 6.8% of Cayman Brac in 2012. It comprises mixed drought-resistant deciduous vegetation on the dolostone karst close to the water table and forms a community alliance by salt tolerant shrubs. The canopy cover is generally greater than 25%, and tree cover generally less than 25% with emergent trees up to 5–6m (16–19ft), and succulents and vines. Areas in depressions in the bluff become flooded seasonally. Dominants include Royen's Tree Cactus, Century Plant, Silver Thatch, Red Birch and Shake Hand together with many species similar to Dry Forest, though of lesser dimensions, including Ironwood, Jasmine, Cherry, Wild Fig, Pepper Cinnamon, Cabbage Tree, Bitter Plum, Whitewood and Broadleaf.

Dry Shrubland originally formed a coastal hedge around the islands comprised by an indigenous community of Sea-grape (few areas remain). This community now includes introduced species Plopnut, West Indian Almond and Coconut Palm, and the invasive Casuarina, and Logwood on Grand Cayman. Xeric shrubland, with a canopy 2–4m tall, occurs in areas of reduced rainfall on the south-eastern and eastern coastal bluffs and along the southern bluff edge on Grand Cayman and Cayman Brac. Dwarf shrubland is prostrate and occurs on the most exposed rocky shores and bluff cliff edges.

Species associated with Dry Shrubland habitat This supports many similar breeding species to Dry Forest habitat although nests are built in emergent trees or tall shrubs. It is the preferred habitat for the endemic Thick-billed Vireo, Bananaquit and Vitelline Warbler, indigenous Northern Mockingbird, Smooth-billed Ani, Gray Kingbird in summer and Yellow-faced Grassquit. This habitat is excellent for passage migrants including Yellow-bellied Sapsucker, two species of cuckoo, flycatchers, vireos, many warblers, Indigo Bunting, and Blue and Rose-breasted Grosbeaks. Turkey Vultures hunt on all three Islands. Brown Boobies nest in dwarf shrubland on the eastern bluff on Cayman Brac.

Mangrove Forest

This evergreen community is seasonally or tidally flooded, at elevations of 0–50cm above mean sea level and is defined by the dominance of one of four species: Red Mangrove, White Mangrove, Black Mangrove and Buttonwood. Buttonwood occurs extensively as a monospecific community, replacing the four- species alliance, in depressions in the dolostone bluff in the eastern half of Grand Cayman, and on Little Cayman.

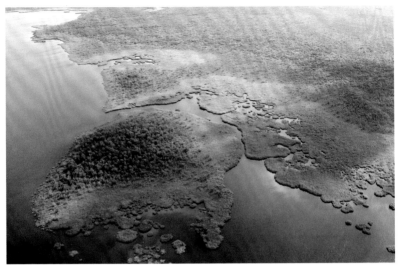

Mangrove Shrubland at the western edge of the 3,400 ha Central Mangrove Wetland bordering North Sound, Grand Cayman

Wet forests, underlain by mangrove peats and Ironshore Formation once covered c. 70% of the land area of Grand Cayman, but are now reduced to 33%, with the Central Mangrove Wetland (3,502 ha; 8,654 acres) the only extensive and ecologically significant area remaining. Wetlands comprise 27% of the land area of Little Cayman of which 16% is mangrove forest. On Cayman Brac, the always small area of wetlands is heavily disturbed and reduced to 0.6% of the land area, two-thirds of which is mangrove forest.

Mangrove Shrubland

Red mangrove is the dominant species, occurring as a low dense fringe around the coastal marine sounds of Grand Cayman (e.g. eastern edge of the North Sound) and Little Cayman (Grape Tree Ponds) and in inland wetlands on both. Monospecific Buttonwood shrubland occurs as linear bands behind the north coast beach ridges on Little Cayman and Cayman Brac, and in the central area of the Central Mangrove Wetland, Grand Cayman.

Species associated with Mangrove Forest habitats Used by breeding West Indian Whistling-Duck and seven species of heron (Least Bittern on Grand Cayman only, Snowy Egret, Tricolored Heron, Little Blue Heron intermittently, Cattle Egret, Yellow-crowned Night-Heron and Green Heron). Landbirds include both woodpeckers, Loggerhead Kingbird, Greater Antillean Grackle, White-crowned Pigeon and Yellow Warbler. Grand Cayman Parrots breed in the remaining Black Mangrove community that forms the boundary between the Central Mangrove Wetland and dry land. Migrants include Merlin, vireos and warblers (especially American Redstart, two species of waterthrush and Common Yellowthroat). Mangrove Shrubland is a very important habitat for migrant warblers.

Saline and brackish lagoons and ponds associated with the mangrove community

These occur behind the beach ridges on all three Islands. Many have direct underground connections to the sea producing delayed tidal effects. Plants include *Batis maritima* and Swamp Fern. The lagoons are seasonal and dry out in spring before the summer rains. On Grand Cayman sites include Meagre Bay, Pease Bay and Fire Station Ponds, Barker's wetland with associated Papagallo, Sea, Vulgunners and Palmetto Ponds, Malportas Pond on the north coast and Collier's Pond in the east. On Little Cayman lagoons almost encircle the island between the coastal beach ridge and the bluff. The major wetlands are Booby Pond, Tarpon Lake, McCoy's Pond, Jackson's and Grape Tree Ponds, Easterly Ponds and Preston Bay westerly wetlands. Remnant mangrove habitat on Cayman Brac in the south-west is associated with the Westerly Ponds, Salt Water Pond and the Marshes.

Species associated with saline lagoons and brackish pond habitats The great majority of wetland birds occur in this habitat, including breeding West Indian Whistling-Duck, Pied-billed Grebe, rails (American Coot and Common Gallinule), Black-necked Stilt, Willet and Least Tern. Almost all migrant wetland species forage on the mud edges or in the water: including herons (including Great Blue Heron, Great Egret, Little Blue Heron, Reddish Egret and Black-crowned Night-Heron), ducks (Blue-winged Teal, American Wigeon, Northern Shoveler, Lesser Scaup), rails (migrant American Coot and Common Gallinule, Sora), plovers (Semipalmated Plover, Black-bellied Plover, Wilson's Plover and Killdeer), up to 18 species of sandpiper and terns (Gull-billed, Common Tern, Forster's and Black Terns); Peregrine Falcons and Merlins hunt waterbirds on the lagoons and Ospreys also do so regularly, although their main prey is fish.

Herbaceous Wetlands
This community lies in shallow depressions in the bluff limestone and is almost completely confined to Grand Cayman, with small areas on Little Cayman and Cayman Brac. The most extensive example is the Salina (275 ha; 680 acres) in north-east Grand Cayman (not easily accessed up to 2012) and a small area at Barker's, West Bay. Co-dominant plants are Buttonwood and Swamp Fern, together with herbs including the endemic *Agalinis kingsii* (Grand Cayman only), Bulrush, Sawgrass, sedges such as *Elocharis* sp. and *Cyperus* sp., and vines. On Little Cayman a seasonal wetland with Sea-purslane and *Salicornia bigelovii* lies in a depression in the central eastern bluff (no trail access) and at Coot Pond. There is a small area at the Marshes, Cayman Brac. Pond weed such as *Ruppia* spp. is taken by West Indian Whistling-Duck. Freshwater wetlands, once an extensive and important habitat on Grand Cayman, are now reduced to <1% of their original extent, occurring at Governor Gore's Pond, Newlands, Mission House Pond, Bodden Town and the west part of the lake at the Botanic Park, Grand Cayman. Fresh water is found in sinkholes in the karst in the dry forest on the three islands, where series of joint fractures in the bluff intersect with the freshwater lens: e.g. the Splits, Cayman Brac.

The Salina, Grand Cayman. The largest herbaceous wetland in the Islands supports wetland birds; landbirds including parrots breed in the Dry Forest (right, not shown).

Fringing reefs enclosing a marine sound attract migrant seabirds, especially gulls and terns, eastern Grand Cayman.

Species associated with Herbaceous Wetland habitats Species preferring freshwater and brackish habitats include breeding Least Bittern and Purple Gallinule, and migrant Black-crowned Night-Heron, White and Glossy Ibis, Sora, Killdeer, Pectoral and Least Sandpipers, Wilson's Snipe and Wilson's Phalarope.

Coastal habitats

These include fringing reefs, the coastline of sandy beaches or Ironshore Formation, and marine bluffs (cliffs).

Species associated with Coastal habitats Fringing reefs, surrounding most of Grand Cayman and Little Cayman (confined to the southwest of Cayman Brac), are backed by shallow marine sounds providing foraging areas for seabirds, waterbirds and shorebirds and attract unusual visitors such as American Flamingo. Antillean Nighthawks breed on the Ironshore and sand beach close to the vegetation line on the three Islands. Ospreys, juvenile Red-footed and Brown Boobies and Brown Pelicans practice diving for fish in the shallow waters. The marine cliffs support breeding pelagic seabirds: White-tailed Tropicbirds breed from January–August on the southern coast at Pedro St. James, Grand Cayman and the north and south bluff face on Cayman Brac; Brown Boobies breed year round on the southern and extreme eastern bluff edge and on ledges and caves in the bluff face on Cayman Brac. Bridled Terns breed on a cay off the north coast of Grand Cayman. Least Terns previously bred on coastal beaches on the three islands but have ceased breeding due to disturbance.

Altered habitats

Extensively man-modified habitats, where primary habitats are mainly replaced with exotic plants, include buildings and roads, golf courses, marinas, dry and flooded marl pits, rough pasture, wasteland, plantations and airports. Urban and urban/littoral habitats remain important for migrant and resident landbirds if some indigenous trees are retained. Dry marl pits (quarries) on Grand Cayman and Cayman Brac support nesting Antillean Nighthawks and Short-eared Owls. On Grand Cayman flooded marl pits and golf courses have wintering rails, herons, ducks and often unusual migrants, such as Black Skimmer and American Avocet. In addition, areas of spoil at construction sites, derived from mangrove wetlands, are important sites for breeding Least Tern, Antillean Nighthawk and Short-eared Owl irregularly. Airport grasslands are good for Cattle Egret, Barn Owl, and migrant Bobolink, Dickcissel, Indigo Bunting, Blue Grosbeak and swallows. Raptors include Northern Harrier. Logwood, introduced in the 18th century to Grand Cayman as a source of dye, has become established, spreading aggressively throughout open tracts of land, especially in damp areas of low salinity. A Logwood/Buttonwood alliance is good for foraging vireos (Thick-billed, White-eyed and Red-eyed), Cuban Bullfinch, Vitelline Warbler and migrant warblers.

Urban development and land cleared for agriculture has led to increased populations of some indigenous birds (mockingbirds, anis, grackles, and feral Rock Pigeon and Monk Parakeet) and to conflict situations, most notably involving egrets and herons at airports (bird strike hazard) and parrots in private and commercially cultivated Mango and Papaya plantations.

THE HISTORY OF ORNITHOLOGY IN THE CAYMAN ISLANDS

The first reference to birds in the Cayman Islands was a brief sentence by a visiting seafarer in 1582. The next came 300 years later, in the 1880s, when the avifauna was first studied by C. B. C. Cory (1857–1921), a wealthy amateur ornithologist and founder member of the American Ornithologists' Union. He described 13 new species of landbirds of which all but one, the Grand Cayman Thrush, were later reclassified as endemic subspecies (Bangs 1916). Cory (1892) produced the first checklist of Cayman Islands' birds naming 55 species, 30 of them breeding. He noted the absence of resident woodpeckers on Cayman Brac and Little Cayman and that the endemic subspecies of the grackle, parrot and thrush on Grand Cayman were replaced on Little Cayman and Cayman Brac by species similar to those on Cuba. The Grand Cayman Thrush (now extinct) remains the only endemic species recognised for the Cayman Islands but, when genetic studies are completed, it seems likely that at least two subspecies will be reclassified as full species. In 1889, Charles J. Maynard first reported the two sulids on Little Cayman and Cayman Brac as a single new species, later named as two: the Red-footed Booby on Little Cayman and the Brown Booby on Cayman Brac. He also conducted the first booby census on Little Cayman.

Avian collectors visited sporadically from 1904–1938, their specimens housed in the museums of Chicago, New York, Washington, Boston (Harvard), London (Tring) and Baton Rouge (Louisiana). Michael J. Nicoll noted the absence of hummingbirds, and that the Vitelline Warbler on Grand Cayman was distinct from that on Little Cayman. Lowe (1911) published the second checklist with 75 species described, of which 40 were breeding.

Brown Booby juvenile begging on the edge of the eastern bluff, Cayman Brac.

Nicoll and Lowe did not see the Grand Cayman Thrush or Baird's Banana-bird (Jamaican Oriole), and the former commented on their scarcity. W. W. Brown in 1911 found the thrush in only two areas of forest on Grand Cayman and believed it to be on the verge of extinction, although he still collected 13 specimens having been 'careful to leave birds to perpetuate the species' (Bangs, 1916)!. Savage English, who lived on Grand Cayman from 1912–1914, did not see the thrush until his third year on the island. David W. Johnston searched extensively for this species in the 1960s and concluded that C. B. Lewis was probably the last person to see the Grand Cayman Thrush in 1938 and that it was now probably extinct. No subsequent searches have been successful (See Appendix 1).

Likewise, Brown in 1911 found the oriole very rare and probably also on the verge of extinction, but still collected 17 individuals. Regrettably, the male Jamaican Oriole that Bond collected on 6 March 1930 was the last specimen of that species. In 1930 Paul Bartsch collected two Northern Mockingbirds on Little Cayman, the first records for the Sister Islands which James Bond believed it did not colonise until the 1960s.

David W. Johnston produced the third checklist citing 151 species of which 45 were breeding and the first ecological study of the avifauna on Grand Cayman and Cayman Brac (Johnston et al.,1971; Johnston 1975). Sebastian Patti noted the absence of the Antillean Grackle on Cayman Brac in 1974, which locals reported lost after the 1933 hurricane. Diamond (1980) made the second census of the Red-footed Booby, Roger Clapp made a third census in 1986. A fourth count by Burton et al. (1999) found c. 5000 pairs or 20,000 birds: a recent census in 2010 suggest that numbers have declined significantly, possibly due to hurricanes in 2005 and 2008. David Wingate made the first observation of the White-tailed Tropicbird breeding on Grand Cayman in 1984. The Cayman Islands' Bird Club was formed in 1989 with Michael Marsden, Trevor Baxter, Peter Fitzgerald, Nancy Norman, Brian Livingstone, Rudi Powery and Peter Davey as founder members. Together with the author they provided data for the fourth checklist of 219 species of which 49 were breeding (Bradley 2000).

The first Cuban Parrot census on Grand Cayman and Cayman Brac was in 1984–1985 (Bradley 1986); the second on the Cayman Brac subspecies (known as the Brac Parrot) was by James Wiley (Wiley et al. 1992), aided by the National Trust and Bird Club. The Nature Conservancy donated 80ha (208 acres) of bluff forest habitat on Cayman Brac to the National Trust in 1993 to create the Brac Parrot Reserve. The National Trust and the Department of the Environment continue to census both subspecies. Orlando Garrido visited Grand Cayman to complete a study of the taxonomy of the polytypic endemic Stripe-headed Tanager, renamed Western Spindalis in the Cayman Islands, Cuba and the Bahamas. Up until 1982, English had been the only naturalist who had lived on any of the Islands for more than a few months. From 1982 the present author gathered data during every month of the year on breeding and migratory birds and their ecology, and in successive years, on each of the three Cayman Islands (Bradley 1994, 1995, 2000, 2009).

BREEDING BIRDS

The current number of species recorded breeding is 50 (in 41 genera in 22 families) of which 20 are waterbirds and 30 are landbirds (Appendices 3–5). The composition of the avifauna is related to the geographical position and small size of the Islands and to their low relief and limited habitat diversity, which determines available niches. Many landbirds that breed on Grand Cayman, Little Cayman and Cayman Brac also breed on Cuba, Jamaica and the American mainland. The closest alliance is with Cuba, with 24 species shared; 19 species are shared with Jamaica, and 18 with the American mainland. There are 18 species endemic to the Caribbean as a whole (Appendix 3) and four of these have established feral breeding populations. There are no surviving Cayman Islands endemic species, but 14 species are represented by 17 endemic subspecies (Appendix 5).

Seabirds, waterbirds, shorebirds (Appendix 3)

Of the 20 species (representing 16 genera in ten families), 16 breed on Grand Cayman, 13 on Little Cayman and 12 on Cayman Brac, and some 16 (in eight families) are resident on one or more of the Islands throughout the year. The West Indian Whistling-Duck is the only endemic duck in the West Indies.

Populations of resident species that breed on the three Islands (West Indian Whistling-Duck, Pied-billed Grebe, Tricolored Heron, Yellow-crowned Night-Heron, Green Heron, American Coot, Common

Snowy Egret flock foraging on saline lagoon, March.

Gallinule, Black-necked Stilt and Willet (the only shorebird) are augmented by migrant populations in winter and on passage. The Snowy Egret breeds on Grand Cayman and Little Cayman. However on Little Cayman regular breeding of waterbirds corresponds with optimum (wet) conditions and some, e.g. Pied-billed Grebe, American Coot, Snowy Egret and Tricolored Heron, do not breed in drought years, and often confine breeding to winter and early spring before wetlands dry out. The Little Blue Heron breeds intermittently on Grand Cayman and Little Cayman and the Purple Gallinule has not bred on Cayman Brac or Little Cayman since 2004.

Seven species that breed regularly show different island distributions: Least Bittern, Purple Gallinule and Bridled Tern on Grand Cayman, White-tailed Tropicbird on Grand Cayman and Cayman Brac, Brown Booby on Cayman Brac, and Red-footed Booby and Magnificent Frigatebird on Little Cayman. Three species are migrants that breed in summer and spend the winter in South America: the Bridled Tern, with a very small population on Grand Cayman, and two species that have shown declines; the Least Tern, which previously bred on all three Islands but only nested on Grand Cayman in 2000–2011, and the White-tailed Tropicbird, whose two populations showed the greatest decline on Cayman Brac. The three resident pelagic seabirds have prolonged breeding seasons: the Red-footed and Brown Boobies have shown declines since the hurricanes in 2008 and 2009, while the frigatebird numbers fluctuate. Sulids and tropicbirds are preyed on by migrant Peregrine Falcons and rats.

Landbirds

Of the 29 species of landbirds (representing 24 genera in 12 families), 27 breed on Grand Cayman, 17 on Little Cayman and 20 on Cayman Brac; 26 are resident, three are summer visitors (Appendix 4). There are no endemic genera, but two genera are endemic to the Greater Antilles, *Spindalis* (Western Spindalis) and *Melopyrrha* (Cuban Bullfinch). There are no surviving endemic species although two did occur: the extinct fossil bullfinch and the recently extinct Grand Cayman Thrush. Several species show a wide habitat tolerance, notably the Bananaquit, which breeds on the three Islands and is frugivorous, nectarivorous and insectivorous.

Restricted range species

Of particular interest and described in detail in the text are four West Indian endemic landbirds of restricted range that have significant populations in the Cayman Islands. The Vitelline Warbler has 97% of its population and two subspecies in the Cayman Islands (3% on Swan Islands). The Cuban Bullfinch is only found on Grand Cayman and Cuba. Ongoing genetic and morphometric studies may result in the bullfinch being reclassified as two endemic species, with *Melopyrrha taylori* on Grand Cayman. The Thick-billed Vireo has a disjunct distribution with

Vitelline Warbler, occurs only on the Cayman Islands and Swan Islands; global conservation status is Near-Threatened.

five separate subspecies: on Grand Cayman, Providencia, Tortue island off Hispaniola, the Bahamas and the Caicos Islands. The Cuban Parrot (Cuban Amazon) occurs as four subspecies, two in the Cayman Islands, and one each in Cuba and the Bahamas. Preliminary genetic studies strongly suggest that the Grand Cayman subspecies may be sufficiently divergent to be reclassified as a separate species.

Endemic subspecies

There are 14 species represented by a total of 17 endemic subspecies, with three species having two subspecies each (Appendix 5). Their distribution is related to island size and ecological constraints. Grand Cayman has 13 endemic subspecies, nine of which are confined only to that island. Cayman Brac has seven and Little Cayman four; one subspecies each occurs only on Little Cayman and Cayman Brac. In the early 20th century subspecies of the Cuban Parrot (Brac subspecies), Loggerhead Kingbird and Thick-billed Vireo bred on Little Cayman (and occasionally all three species fly over from Cayman Brac to forage) and the Greater Antillean Grackle bred on Cayman Brac.

The Cuban Parrot *Amazona leucocephala* is classified as Near-Threatened. The Grand Cayman subspecies *caymanensis* numbers about 2,500–3,000 birds and while the greater part of the population occurs in the centre and east, small flocks are often observed in urban George Town and the West Bay peninsula. The Cayman Brac subspecies *hesterna* combines a small population size with the smallest range of any amazon parrot, factors which, together with on-going habitat loss and low breeding success, are causing grave concern regarding its future status. Pairs do not nest every year and long-term studies are needed to assess whether the >100 pairs or so that nest each season can sustain the population and also whether low breeding success is due to an ageing population, lack of suitable nest cavities and/or constraints on food availability. This subspecies is at a further disadvantage when selecting nest sites since there are no breeding woodpeckers on Cayman Brac to initiate cavity construction. In Grand Cayman, parrots frequently use old woodpecker nest holes. A management plan for the Brac Parrot will almost certainly require intervention to provide more protected habitat, the possible provision of artificial nests and ultimately, if all else fails, a captive breeding programme.

Breeding distribution

The majority of landbirds have breeding peaks in spring to early summer and five species (White-winged Dove, Common Ground-Dove, Smooth-billed Ani, Bananaquit and Yellow-faced Grassquit) breed in all months on Grand Cayman. The majority of indigenous resident species breed on all three islands except La Sagra's Flycatcher which is confined to Grand Cayman. The White-crowned Pigeon and White-winged Dove are mainly migratory and their numbers are lowest in winter. Populations of the Smooth-billed Ani and Northern Mockingbird increased greatly following the three hurricanes and two tropical storms that occurred during 2004–09 which, alongside development, expanded the amounts of disturbed and cleared habitat. Concomitantly, populations of the pigeon and the Zenaida Dove have declined. A small population of the Short-eared Owl breeds on the three Islands with nesting observed on the ground at all three airports; also immature birds on post-breeding dispersal most likely come from Cuba, where the population has increased sharply. The Loggerhead Kingbird populations on Grand Cayman and Cayman Brac have declined sharply following the 2004 and 2008 hurricanes. Three species are austral migrants from South America: the Antillean Nighthawk and Gray Kingbird breed on the three Islands but the Black-whiskered Vireo breeds only on Cayman Brac and Little Cayman. The Shiny Cowbird is making persistent attempts to colonise Grand Cayman, having colonised Cuba in 1982, Jamaica in 1992 and the Bahamas in 1994. There have been two records of single pairs breeding. Should the cowbird become established it will present a major threat to endemic subspecies of warblers and vireos. Its negative impact on the endemic Puerto Rican Vireo is well documented.

Moults

A bird in the field may not look exactly like the photograph in the book. It is important to recognise that birds' feathers undergo changes throughout the year, from worn and faded to bright and new following a full moult. Many species assume these changes gradually. Others have completely different plumages in non-breeding and breeding seasons, and may also show intermediate stages while in transition between the two, e.g. Black-bellied Plover. A full or partial moult to replace worn feathers usually occurs twice a year including the post-breeding moult, which may take place before migration, at migration stops or on arrival in the wintering grounds. The moult from immature to adult can result in a brighter but similar plumage, e.g. Ovenbird, or one distinct from the immature phase, e.g. Little Blue Heron. Most juveniles have similar but duller plumages to adults with paler feather tips and fringes, as in doves and many shorebirds, while others have distinctive field marks, e.g. Western and Least Sandpipers have short-lived bright cinnamon feathers on the face and scapulars, an important identification tool in early fall. In most passerines, juvenile plumage lasts until late summer, the first-winter plumage follows and remains until the first spring when it is replaced by the first-summer plumage, which is often similar to the adult breeding plumage. The first adult non-breeding plumage is assumed through the second fall and winter and replaced by adult breeding plumage in spring thereafter. In some species replacement occurs within a shorter period, e.g. Bananaquit and Yellow-faced Grassquit. In some seabirds, e.g. gulls, there are distinct juvenile, first-, second-, third- and sometimes fourth-year plumages before adulthood. Other species assume adult plumage gradually, e.g. Red-footed Booby over three years.

Thus between fall and spring different individuals of a single species may exhibit juvenile, immature, male, female, breeding and non-breeding plumages as well as transitional plumages. It takes many years of expertise to distinguish between the dull fall plumages of non-breeding and immature birds, especially some shorebirds and warblers. In fall and winter in the islands, the majority of migrants are in non-breeding and immature plumages, with first-year warblers and shorebirds much in evidence. However in late spring, migrants display adult breeding plumage which is dramatically different from the winter or non-breeding plumage in the cases of some shorebirds and warblers.

MIGRATION

The geographical position of the Cayman Islands, at the western edge of the Greater Antilles close to the boundary between the Caribbean Sea and the Gulf of Mexico and midway between the North and South American continents, accounts for the greater part of its avifauna (c. 80%) occurring as non-breeding migrants. This makes bird watching in the Cayman Islands an exciting challenge in fall and spring as many species that pass through the western Caribbean and cross the Gulf of Mexico to their main destination in Middle America can be expected to occur there. The paucity of records relates directly to the paucity of observers. There has been no long-term ringing study of migrants, so their status is based on systematic and regular observations and on counts made at sample sites on all three Islands, and on data gathered from the literature and museum specimens.

The majority of migrants breed in the Nearctic, from where all or some of their populations migrate to winter in the Neotropics (the West Indies, Middle and South America) where they become part of the tropical ecosystem for up to nine months of the year. Migrants reach the Cayman Islands in fall via three main routes: the Mississippi flyway and across the Gulf of Mexico, the Atlantic flyway via

Florida and the Greater Antilles or the Atlantic seaboard through the Bahamas into the Antilles. Some are known to travel to the western Caribbean from the Yucatan. The majority of migrant landbirds are insectivores on their breeding grounds (swifts, flycatchers, vireos and warblers) but many tend towards omnivory when in the tropics, taking an increased amount of fruit and seeds. Thus they exert a considerable influence on the fruiting and flowering of tropical plants.

Conservation of neotropical migrants requires an understanding of migration connectivity: the ecological conditions encountered on the migration route, on the wintering grounds, at stopover sites along the migration route, and on the breeding grounds. Migrants must also overcome conditions during the migration flight when wind direction and tropical storms and hurricanes are relevant. The increasingly rapid loss of tropical forest is known to be partly responsible for the long-term decline in the total number of migrant songbirds. It leads to the use of sub-optimal habitats in the wintering grounds resulting in birds in poor condition arriving late on the northern breeding grounds with a reduced ability to compete for mates and nest sites.

More species are recorded in the Islands, in greater abundance, in fall than in spring due to more favourable winds and because many species return northwards via the Middle American land route, thus bypassing the western Caribbean. It should be noted that there are wide yearly fluctuations in abundance in all migrant populations in the Cayman Islands, linked to both local climatic conditions and regional weather patterns. All migrants are more numerous in the fall in years when a wet early spring is followed by good summer rains; after dry winters, spring migrants are fewer in number and spend shorter stopover periods, or even overfly the Islands without stopping. Many common migrants in Cuba are also common on the Cayman Islands. However recent data from Cuba suggests that several species, previously thought to migrate further west, regularly extend their flight eastwards to cross the western Caribbean and a few winter in low numbers. It has been surmised that the high frequency of out-of-range species in the region might be because new winter territories are being tested. Several species previously thought to be rare on Cuba are now known to winter there regularly and are appearing on the Cayman Islands increasingly often.

The majority of migrants arrive between late August and November and leave in May. Stragglers continue to arrive in December and at least one, the Yellow-rumped Warbler, may arrive up to early January. The Yellow-rumped Warbler and Cedar Waxwing are irruptive migrants, exceptionally abundant in some years, linked to fruiting phenology, and rare or absent in others. The earliest wintering migrants leave in February, and the latest in May. Those on spring passage from further south pass through the Islands from January until early June and, because of the proximity of North America to the Cayman Islands, the earliest southbound migrants in June may overlap with the latest northbound birds, so that some migration occurs year-round.

Adult Black-crowned Night-Heron eating Tilapia on Gore's Pond, Grand Cayman.

Regular migrants

Well over half of the non-breeding migrants are recorded regularly with a variety of distribution patterns. Some species are recorded in all months, either as short-stay passage migrants (Wilson's Plover, Barn Swallow, Laughing Gull) or where a few immature or non-breeding adults remain throughout the summer after the majority have migrated (Brown Pelican, Glossy Ibis, Great Egret, Black-crowned Night-Heron, Semipalmated Sandpiper and Ruddy Turnstone). For others (most shorebirds, warblers, Indigo Bunting) the greatest numbers occur on passage with only a small number overwintering. For species that regularly winter in the Greater Antilles (Spotted Sandpiper, American Redstart) a late spring passage from further south is noticeable after the wintering birds have left. Other species (Sora, Yellow-bellied Sapsucker, Palm Warbler) arrive in the fall, winter in the Islands and return north in the spring, and no separate through passage is observed. Several species, such as the White-rumped Sandpiper and Blackpoll Warbler, only occur on passage. The latter occurs in spring having made a fall trans-oceanic migration down the eastern Atlantic seaboard via the Lesser Antilles to South America. Some species makes a trans-Gulf migration to winter in Middle America and pass through the Cayman Islands more frequently in fall (Red-eyed Vireo, Swainson's Thrush, Blue-winged Warbler, Chestnut-sided Warbler).

While short-term changes in abundance of migrants are to be expected, longer term trends (over 30 years) have also been observed. In some cases this reflects changing status on the Nearctic breeding grounds. For example, the Swallow-tailed Kite was once rare in passage but is now regular in large flocks, due to increased populations on its northern breeding grounds. The Peregrine Falcon, previously very uncommon on passage in the 1980s, is now regular in both winter and passage reflecting conservation efforts that restored healthy populations to North America. This latter success, however, has come at a cost to our booby, tropicbird and heron populations which provide the Peregrine's main prey, and has certainly contributed to the decline of the Brown Booby on Cayman Brac. Other species (Blue-grey Gnatcatcher and Savannah Sparrow), that were regular on the islands in the 1980s, have declined in numbers in North America and have also become rare here.

Warblers represent the majority of North American passerine migrants with 34 of the 50 known species recorded. The most common are Northern Parula, Palm Warbler, American Redstart (now breeding on Cuba), Black-throated Blue Warbler, Prairie Warbler, Black-and-white Warbler, Northern Waterthrush, Ovenbird, Yellow-throated Warbler, Cape May Warbler, Worm-eating Warbler and Yellow-rumped Warbler. All, except the terrestrial Palm Warbler, select habitats occupied by the two resident warblers. The high density of migrants in winter almost certainly has an impact on the resident species and may be the reason why only two warbler species are resident here, as in Cuba.

Vagrants and rare visitors

A large number (around 50 species) of migrants are rare visitors or vagrants with fewer than ten records each, reflecting the small size of the Island group and its geographical position (Appendix 6). Vagrants include North American migrants that are regular in the Greater Antilles, e.g. Cuba, between August and May (Wood Thrush); species that breed in the Greater Antilles (Eastern Meadowlark, Black-capped Petrel, American Bittern), and species that only occur after the passage of hurricanes and tropical storms in fall or of a cold front in spring, when they are blown off their usual migration route (Buff-breasted Sandpiper and predominately Middle and South American species such as Fork-tailed Flycatcher and Tropical Kingbird).

Buff-breasted Sandpiper transitioning to breeding plumage, vagrant, May.

Introduced and Colonised Species

The growth of the cage bird trade on the Cayman Islands has led to over 60 species having been bred in captivity on Grand Cayman. Many escape and others are released deliberately and, while almost all are either recaptured or perish, a few manage to establish feral breeding populations, e.g. Rock Pigeon on Grand Cayman and Cayman Brac, and Monk Parakeet. The latest arrival, House Sparrow on Grand Cayman, may have colonised from Cuba or Jamaica but is more likely to have come in a ship container from Florida. Exotic species establishing breeding populations are one of the greatest threats to birds on islands world-wide as they compete with endemic species for nest sites and food.

Three other species had established feral breeding populations in the 1990s: Rose-ringed Parakeet and Red-masked Conure appear to have succumbed to the effects of the 2004 hurricane. The status of the Yellow-crowned Parrot is uncertain, although breeding is suspected. Occasional individual escapees: Ringed Turtle-Dove, Sulphur-crested Cockatoo, Budgerigar, Cockatiel and Tricolored Munia, are observed but none have become established.

CONSERVATION

The Cayman Islands, together with the Galapagos, were unique in the Americas in being uninhabited from pre-Columbian times until the arrival of the first permanent settlers in the mid-18th century. But even before settlement, between the mid 17th to mid-18th century, the impact of visitors had resulted in the largest forest trees being almost completely logged and the once seemingly inexhaustible supplies of Green Turtles being exhausted. By the end of the 19th century, large tracts of hardwood forest had been cleared for imported Logwood and cotton plantations. The result, by the mid 20th century, was the local extinction of five bird, two mammal and one reptile species.

Conservation Legislation

The Cayman Islands Government (CIG) is a signatory to a number of international conservation agreements including the Ramsar Convention, Convention on International Trade in Endangered Species (CITES), Cartegena Convention and Spaw Protocol, Bonn Convention and the Biodiversity Convention.

The endangered endemic *Cyclura lewisi* has been released in the National Trust's new Collier's Wilderness Reserve, Grand Cayman.

The Animals Law (2003 revision) is the only means available to the CIG to declare protected areas or species. Under this law, all species of birds are protected, except the White-crowned Pigeon, White-winged Dove and Blue-winged Teal, which may be hunted from 1 August to 31 January. A revision has allowed the removal of the invasive Green Iguana *Iguana iguana* from the protected species list, while protecting the endangered endemic Blue Iguana *Cyclura lewisi*. The Marine Conservation (Marine Parks) regulations (2007 Revision) protect inshore Marine Parks and the eastern boundary of the Central Mangrove Wetland, North Sound, Grand Cayman.

The National Trust for the Cayman Islands Law (2003 Revision) is a conservation law that protects the Trust's land holding and allows land to be declared inalienable.

Government Conservation Policy A CIG Ministry is responsible for environmental issues through the Department of the Environment. There are comprehensive Marine Park Laws and Regulations regarding closed seasons, catch limits, marine parks and replenishment zones which are strictly enforced by Marine Officers of DOE as a matter of Government policy. There is no such policy or laws for terrestrial issues: no National Parks or comprehensive Development Plans that include sustainable land use.

The proposed National Conservation Law, in draft for 12 years, would allow the Cayman Islands Government to meet its obligations under these conventions and provide greater flexibility and powers to create and manage protected areas and conserve species. Sadly, up to 2012, successive Governments have refused to pass the law leaving a large gap in the ability to implement proper conservation planning and environmental management. In addition, the requirement for developers to undertake Environmental Impact Assessments is not enshrined in any legislation. To date the law seems unlikely to reach the statute books without much modification.

Protected areas

In 2012, protected areas totalled 7.74% of the combined land area of the three Cayman Islands. The National Trust owns and protects 1,292ha (3,193 acres) of dry and wetland areas on the three Islands.

Grand Cayman The Crown (CIG) owns 541ha (1,339 acres) of land in the 3,502ha (8,654 acres) Central Mangrove Wetland, Grand Cayman and 690ha (1,707 acres) as the Marine Park Environmental Zone, which include coastal water and mangrove edge along the western boundary. The great majority of

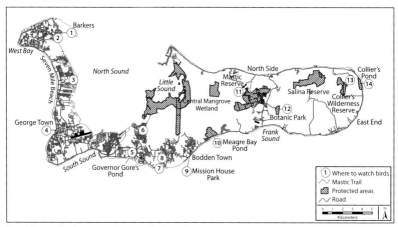

Grand Cayman showing protected areas and main roads.

1. Barker's Peninsula
2. Morgan's Harbour marina
3. Golf course at Safe Haven
4. George Town Harbour beach
5. Governor Gore's Pond
6. North Sound Estates
7. Pedro bluff cliffs
8. Agricultural Grounds
9. Mission House Pond, Bodden Town
10. Meagre Bay Pond
11. Mastic Trail and Mastic Reserve
12. Botanic Park
13. Collier's Wilderness Reserve
14. Collier's Pond

the wetland remains in private ownership. There are two Crown Animal Sanctuaries: Meagre Bay Pond on the southern boundary of the wetland, and Collier's Pond in the eastern districts, but the mangrove buffer surrounding each pond is privately owned allowing development up to the water's edge.

The National Trust owns 309ha (765 acres) of the Central Mangrove Wetland, 275ha (680 acres) of the Salina Reserve, and shares 50ha (130 acres) of the Botanic Park jointly with CIG. The Trust owns 297ha (734 acres) of the Mastic Reserve, the remaining significant area of dry forest on Grand Cayman, with further purchase intended as land and funds become available. The Governor Michael Gore Bird Sanctuary and the Mission House Pond, each less than one ha, are the only protected freshwater sites among the few scattered remnants on Grand Cayman. The CIG has leased 76ha (190 acres) of dry shrubland to the National Trust for 99 years to form the Collier's Wilderness Reserve in the eastern districts as a sanctuary for the endangered and endemic Blue Iguana.

Cayman Brac Salt Water Pond, an Animal Sanctuary owned by CIG, is to be deregulated (for a marina) and the other two coastal lagoons have already been deregulated because of their close proximity to the international airport.

The National Trust protects 115ha (286 acres) of Dry Forest on the bluff as a Parrot Reserve, and 11ha (26.5 acres) of Dry Forest at the Splits. All other land is privately owned.

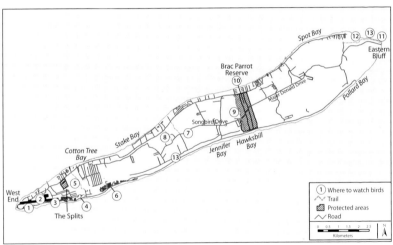

Cayman Brac showing protected areas and main roads.

1. Westerly Ponds	6. The Marshes	11. Eastern Bluff
2. Airport runway and grassland	7. Songbird Drive	12. Lighthouse Walk
3. Coastal woodland opposite an abandoned hotel	8. Deadman's Point Bluff road	13. The bluff face on the north and south coasts
4. Salt Water Pond	9. Bight Road Dry Forest trail	
5. Salt Water Pond Walk	10. National Trust Parrot Reserve and Nature Trail	

Little Cayman Wetlands comprise 27% of the total area of the Island of which 85% are owned by the CIG and, as yet, have no protected status. The National Trust owns Booby Pond Reserve, a Ramsar Site (Wetland of International Importance), covering 132ha (326 acres) of lagoon, Mangrove, Dry Forest and Shrubland. It protects the large Red-footed Booby colony, a Magnificent Frigatebird colony, a heronry, migrant waterbirds and endemic landbirds. The National Trust owns 14ha (33acres) of Dry Shrubland at the western end to protect the Sister Islands Rock Iguana.

Following private land claims, the CIG has returned some Crown wetlands to private ownership, and another 25% of the wetlands are subject to further such claims, a cause of grave concern.

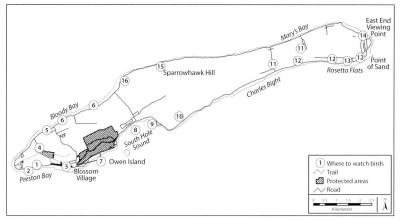

Little Cayman showing protected areas and main roads.

1. Preston Bay westerly ponds
2. Preston Bay coastal woodland and dry shrubland
3. Airstrip
4. Salt Rock Nature Trail
5. McCoy's Pond
6. Coastal Mahogany forest
7. Booby Pond Reserve
8. South Hole Sound and Owen Island
9. Kingston Bight
10. Tarpon Lake
11. Central bluff forest
12. Easterly Ponds
13. Coot Pond
14. East End Viewing area
15. Grape Tree Ponds
16. Jackson's Pond

Protected Dry Forest in the National Trust-owned Mastic Reserve, Grand Cayman.

Threats to biodiversity and the avifauna

The Caribbean is one of the richest areas of endemism in the world, leading international conservation organisations to call for resource planning as a matter of urgency in this region of small island states. Most governments are confronted with rapid human population growth and increasing demands for improved standards of living amid a declining GDP. It is apparent that uncontrolled development is not the hoped for economic solution for small islands because it threatens to outstrip the limited sustainable natural resources. Caribbean islands at the lowest end of the economic scale have raised concern as doomsday scenarios unfold: exhausted resources, water and fuel shortages, deforestation and erosion, child malnutrition and mortality, crime and civil unrest.

Lack of sustainable land use planning The success of Grand Cayman's economy from the 1970s to 2004 led to an escalation in the commercial value of land, a booming financial industry and a population that increased from 18,000 to 55,000. To date, this has resulted in reluctance on the part of successive Governments to introduce environmental planning legislation which, it is feared, would have an impact on economic growth. Meanwhile the damaging effects of such rapid growth are already evident: pollution of groundwater, coral reef degradation, dependence on food and fuel imports, loss and fragmentation of forest, and escalating demands for dredged marl to fill the cleared wetlands.

At present there is no Government legislation to create Development Plans for the three Islands that would identify environmentally sensitive areas and promote sustainable land use. As a matter of urgency, CIG should introduce Development Plans for each Island to include an environmental overlay, despite this issue remaining a political hot potato. (The 37-year battle seeking to designate the Central Mangrove Wetland as an environmental zone in a proposed Development Plan failed again in 1999 and 2003, causing such controversy with landowners that the issue has been postponed indefinitely.)

Lack of Protected Areas There is no CIG legislation to create a land-based system of National Parks, another matter of urgency. Land is so expensive that it is impossible for the National Trust to continue to be the only organisation to purchase land to protect areas identified as essential to preserve the Islands' biodiversity. It is now critical that the CIG intervenes to use parts of the Environmental Fund to purchase land to preserve habitats and species and benefit future generations of Caymanians.

BirdLife International, through a system of Important Bird Areas (IBAs), seeks to identify and protect a network of globally important sites critical for bird conservation. The Cayman Islands have had ten sites accepted as IBAs within the Caribbean Area and these afford a valid basis for a system of National Parks. The sites hold one bird species of global conservation concern, two species of Near-Threatened status, six species of Restricted Range in the Caribbean, and one species of seabird in sufficient numbers to qualify its site as an IBA. Having accepted this initiative, the CIG is now responsible for working towards securing full legal protection for all ten sites. Five IBA sites have some legal protection afforded by CIG and the National Trust (NT). Five sites have no legal protection (three areas of Dry Forest in the eastern districts of Grand Cayman, Dry Forest and Shrubland at Sparrowhawk Hill and all the wetlands owned by CIG on Little Cayman). All should become National Parks.

Loss of biodiversity If the present situation continues, avian diversity, indeed biodiversity, will not be sustained and species will be lost in the Cayman Islands. Loss and fragmentation of avian habitat is known to have profound cumulative implications for breeding birds and overwintering and stopover migrants in the region. On Grand Cayman the populations of five species of endemic landbirds have sharply declined west of Savannah and throughout the West Bay Peninsula, where virtually all the primary wetland and terrestrial habitats have been lost or fragmented. The endemics have been replaced by a few species able to adapt to urban areas. A completed Darwin Initiative Project has produced a National Biodiversity Action Plan (NBAP) to include habitat maps as well as species and habitat action plans. These indicate current conservation status, known threats and required legislation and conservation management needs. Unfortunately the NBAP cannot be properly implemented without the modern conservation framework envisaged by the draft Conservation Law.

Loss of forest, shrubland and wetlands, as well as fragmentation, has a major impact on diversity. On Cayman Brac the rapid loss of forest is a recent phenomenon as new roads and houses are being constructed throughout the bluff. Fragmentation of forest increases predation by allowing access to rats, cats and dogs and is a major threat to all birds. Such habitat changes have profound implications for the Brac Parrot with a breeding population of >100 pairs. The bluff forest protected as the Brac Parrot Reserve has insufficient breeding habitat to sustain this Near-Threatened endemic subspecies, which depends on the entire Dry Forest for breeding and foraging.

Dry Shrubland habitat on Grand Cayman and Cayman Brac has been reduced to 7% of the land area of each island, making this declining habitat in need of prompt protection. On Little Cayman a proposed new airport, which would have affected the Red-footed Booby colony, has been shelved due to the present economic climate. However, large development projects have opened new roads throughout pristine areas of the eastern half of Little Cayman. These present the first major threat to the interior Dry Shrubland and forest of this fragile Island. Freshwater wetlands on Grand Cayman have been almost completely lost to development.

Illegal hunting continues to threaten species of global conservation concern, including the two subspecies of parrot which continue to be shot. Unfortunately, no prosecutions have resulted from any recent illegal actions. Illegal shooting of the West Indian Whistling-Duck has declined in the 21st century. Legal hunting has also declined but remains a real threat to the declining populations of the White-crowned Pigeon (which remains on the list of game species) as birds are shot while young are still in the nest.

Introduced species are one of the greatest threats to endemic island forms. The uncontrolled nature of the cagebird trade is another major concern because potential pest species are permitted to be imported and sold, many escape or are released on a regular basis on Grand Cayman and a few manage to breed. Already, four introduced bird species have established breeding populations. The Green Iguana on Grand Cayman is an example of how an introduced species with no natural predators can become a major pest within a few years and evidence is mounting that this 'vegetarian' also takes birds' eggs and fledglings.

Climate change Changes in the climate are likely to have grave consequences for some of the Caribbean's birds. One of the undisputed facts is that sea temperature rise and increased levels of dissolved atmospheric carbon dioxide will increase the acidification of the ocean. Acidification has far-reaching ecological consequences for marine life, including damaging calcium uptake which affects growth rates of coral reefs and of the crustaceans that form the bulk of the zooplankton. This impact continues up the food chain to other major consumers, such as fish and squid, reducing their reproductive success and thus causing population declines. Reduced fish stocks, as well as affecting humans, will affect the large seabird colonies of the Cayman Islands and the Caribbean region, which are already showing declines.

Sea level rise as a result of global warming is a threat to the low-elevation Cayman Islands as saltwater inundation will alter the extensive mangrove ecosystem, causing coastal erosion. This system supports many breeding and wintering birds which would be displaced. Another predicted impact is for stronger hurricanes in the region with an increased frequency of Category 3–5 tropical systems.

Conservation efforts

The formation of the National Trust in 1987 was the most important event for conservation in the Islands. In the subsequent 25 years, the Trust has gained local and international support for its efforts to create understanding and awareness of the islands' biodiversity and the necessity of protecting sensitive ecosystems as well as purchasing land for preserving biodiversity. The Cayman Islands Government has given parcels of Crown land to the National Trust on the each of three islands.

The major conservation successes for the National Trust are the protection and stabilisation of the West Indian Whistling-Duck and the Grand Cayman Parrot populations and providing reserves for the parrot on Grand Cayman and Cayman Brac and for the Red-footed Booby on Little Cayman. On Grand Cayman the establishment of the Mastic Reserve has been critically important in protecting much of the remaining pristine Dry Forest, while the Collier's Wilderness Reserve protects the shrinking Dry Shrubland habitat and provides sanctuary for the critically endangered Blue Iguana, and the protection of remaining herbaceous wetlands on Grand Cayman at the Salina, Governor Gore's Pond and the Mission House Pond has been achieved. The environmental programme aims to protect entire ecosystems so that all species of plant and animal benefit. The Trust also runs education programmes to raise awareness among generations of the Nation's schoolchildren.

International organisations have supported the Trust's efforts and provided grants and assistance for conservation biology. They include The Nature Conservancy, Rare Center for Tropical Conservation, National Fish and Wildlife Foundation, Durrell Wildlife Conservation Trust, the Royal Society for the Protection of Birds, World Wide Fund for Nature, and American Birding Conservation, and UK Foreign and Commonwealth Office. Many other organisations and institutions have shared expertise and personnel including the West Indian Whistling-Duck Working-Group (WIWDWG), a branch of the Society of Caribbean Ornithology.

There is growing support among ornithologists for the view that many avian endemic island subspecies in the Caribbean Region would be recognised as single island endemics if subjected to genetic analysis and closer morphometric study. This would greatly aid conservation efforts since it is much easier to arouse interest in a single (endemic) species than in a form that is just one of several subspecies. BirdLife International has now acknowledged the importance of avian endemic subspecies on small islands. In the Cayman Islands three species qualify for such analyses: the Cuban Parrot, Cuban Bullfinch and Vitelline Warbler.

Further monitoring of endemic and resident birds is essential to assess risk indicators, identified by IUCN (2009) as pointers to endangerment, and is especially relevant on small, isolated, low-lying archipelagos such as the Cayman Islands. Even without habitat loss, our endemic birds face a precarious existence due to small range sizes (Vitelline Warbler, Cuban Bullfinch), small population sizes (Cayman Brac Parrot, Greater Antillean Grackle on Little Cayman), some population declines (five endemic subspecies in western Grand Cayman) and the need for specific climax habitats (the parrot as a hole nester is dependent on mature forest trees). Knowledge of these limiting parameters, combined with securing and preserving habitat, is the only route to ensuring the future of the Cayman Islands' birds.

There is a need for a new partnership between the CIG, the National Trust, developers and land owners to bring about a recognition and acceptance of the intrinsic value to human well-being of keeping the remaining wetlands and forests intact and to conserving their birds, other animals and plants.

The only West Indian endemic duck enjoys a nature-tourist viewing platform, Little Cayman.

Herons, shorebirds, gulls and terns occur in the mangrove wetlands and beach on the north shore (right), Barker's Peninsula, Grand Cayman.

WHERE TO WATCH BIRDS IN THE CAYMAN ISLANDS

Please note that some areas mentioned in the Introduction are not included in 'Where to watch birds'. This is because the land is inaccessible to observers. Either it is privately owned or it is considered unsafe due to an absence of trails through the nearly impenetrable vegetation and karstic rock. Below are listed the best places to find birds at the time of writing. However it is important to note that ONLY areas marked as PROTECTED are not subject to development.

GRAND CAYMAN (see map on page 29)

1. Barker's Peninsula
Follow signs to West Bay and Papagallo Restaurant. Continue on the dirt road into an extensive wetland of mangrove with saline lagoons and ponds intersected by canals, and including the northern coastline. Good for herons, shorebirds, Barn Owl, Osprey, Peregrine Falcon, Merlin, gulls and terns, and for warblers in passage in the coastal sea-grape.

2. Morgan's Harbour marina
Fishermen cleaning fish attract frigatebirds and migrant pelicans, gulls and terns.

3. Golf course at Safe Haven
This area may become closed to observers. Excellent for breeding West Indian Whistling-Duck, Peregrine Falcon, Merlin, and rare shorebirds, ducks, gulls and terns, and Short-eared Owl.

4. George Town Harbour beach
Fishermen cleaning fish bring frigatebirds and migrant pelicans, gulls and terns, shorebirds.

5. Governor Gore's Pond
PROTECTED. Turn off main road (Shamrock Drive) into Spotts-Newlands Road. Follow signs to Bird

Governor Gore's Pond. An important freshwater site for waterbirds and for warblers in the surrounding shrubland, Grand Cayman.

Sanctuary. Turn right into Pennsylvania Avenue. Freshwater pond with herons (especially migrant Black-crowned Night-Heron and breeding Least Bittern), rails including Purple Gallinule and Sora, breeding Pied-billed Grebe and Black-necked Stilt, Belted Kingfisher, and migrant shorebirds including snipe, swallows and many warblers in the vegetation east of the pond.

6. North Sound Estates
UNDERGOING DEVELOPMENT. Take Hirst Road and turn right into North Sound Estates. On Rackley Boulevard turn right into Southward Drive, becoming Winward Road around eastern border of the estates. Just past Nadine Road turn right to enter a path along the canal dykes through the edge of the Central Mangrove Wetland. In winter large feeding aggregates of Snowy Egrets and Tricolored Herons occur; also breeding White-crowned Pigeon, woodpeckers, parrots, Barn Owl, Antillean Nighthawk in summer, Loggerhead Kingbird, Yucatan Vireo, Yellow Warbler and breeding colonies of Greater Antillean Grackles. Also migrant Sora, cuckoos, Merlin, American Kestrel and migrant warblers.

7. Pedro bluff cliffs
Follow signs to the historic Pedro St James site and walk to the bluff cliffs on the south coast. A small colony of White-tailed Tropicbirds breed in holes in the bluff close to the sea, January–June.

8. Agricultural Grounds
The Lower Valley forest is excellent for resident breeding birds and migrant flycatchers, thrushes, warblers and tanagers; check cleared pasture for Short-eared Owl, swallows, Bobolink, sparrows and buntings.

9. Mission House Pond, Bodden Town
PROTECTED. Follow signs on Gun Square Road to the Mission House. At the freshwater wetland, look for herons, West Indian Whistling-Duck, migrant ducks, Sora, Purple Gallinule, shorebirds, warblers, flycatchers and vireos in the shrubland.

10. Meagre Bay Pond
PROTECTED. Large, saline lagoon on the inland side of the main road east of Bodden Town. Good for breeding West Indian Whistling-Duck and Least Tern (April to August) and migrant herons, ducks (especially Lesser Scaup and Ring-billed Duck), rafts of migrant American Coots and Pied-billed Grebes (a few

pairs of each breed), shorebirds, Osprey, Merlin, pelicans and an occasional flamingo.

11. Mastic Trail and Mastic Reserve

PROTECTED. Follow signs from the Frank Sound Road, crossing the island north-south. Turn left along a track to the National Trust car park. Take a 5km hike through mangrove and ancient dry forest. Open to the public but booking the National Trust guided walk (Tel. 345-749-1121) gives the best understanding of Cayman's birds, endemic plants and wildlife, and local history. Good for all endemics especially parrots, Caribbean Dove, Yucatan Vireo and Cuban Bullfinch. Also excellent for

The Cayman Islands have a sizable population of West Indian Whistling-Duck, whose Global Conservation Status is Vulnerable.

migrant thrushes, flycatchers, vireos and warblers. The northern entrance to the Mastic Trail on the north coast is excellent for migrant warblers.

12. Botanic Park

PROTECTED. Off Frank Sound Road. Entrance fee. Excellent for breeding West Indian Whistling-Duck, rails including breeding Purple Gallinule, and Least Bittern on the lake. In the forest and shrubland look for Caribbean Dove, parrots, two vireos, Vitelline Warbler and Cuban Bullfinch. Also good for migrant landbirds in passage and Ruby-throated Hummingbird.

13. Collier's Wilderness Reserve

PROTECTED. National Trust Reserve for the endemic and endangered Blue Iguana. Good for endemic landbirds, Vitelline Warbler, Thick-billed Vireo and Cuban Bullfinch.

14. Collier's Pond

PROTECTED. Excellent for all waterbirds, including migrant plovers and sandpipers.

Migrant herons and shorebirds visit Collier's Pond and shoreline, eastern Grand Cayman; note the interior Dry Shrubland.

Western Cayman Brac showing the wetlands and international airport.

CAYMAN BRAC (see map on page 30)

On Cayman Brac all sites are marked on the Department of Tourism's nature tourism brochure.

1. Westerly Ponds

Viewing areas. Almost all wetland birds in the Cayman Islands have been observed at these two hypersaline lagoons, including West Indian Whistling-Duck and many vagrants.

2. Airport runway and grassland

Ospreys and Peregrine Falcons frequently prey on Cattle Egrets and shorebirds; Short-eared Owls breed at the east end of the runway, also good for migrant Purple Gallinule, Buff-breasted Sandpiper, swallows and Bobolink.

3. Coastal woodland beside an abandoned hotel south of the airport

Migrant flycatchers, warblers and tanagers. Antillean Nighthawks nest, and Barn Owls hunt, on the coastal Ironshore.

4. Salt Water Pond

At junction of Southside Road and Gerrard Smith Road. Look for Least Tern from April to August, and herons.

5. Salt Water Pond Walk

Steps at Rebecca's Cave lead to a woodland trail and boardwalk extending halfway across the island. Breeding Zenaida Dove, Common Ground-Dove, Caribbean Elaenia, Bananaquit, Vitelline Warbler and some migrant warblers.

6. The Marshes

Opposite Public beach, an extensive wetland on pavement Ironshore has resident and wintering herons, duck, rails, ibis and shorebirds including Wilson's Snipe.

7. Songbird Drive

Forest, agricultural and urban. In the mid-bluff east off Ashton Reid drive. Good for feeding White-crowned Pigeon, Zenaida Dove, Cayman Brac Parrot, Black-whiskered Vireo, Caribbean Elaenia and Gray Kingbird (in summer) when trees are fruiting; also Barn Owl. Centennial Park on Aston Rutty Drive has remnant forest and is also good for landbirds.

The Marshes Herbaceous Wetland below the bluff, southwest coast, Cayman Brac.

8. Deadman's Point Bluff road

Up steps on the north coast or enter the trail from Songbird Drive. White-crowned Pigeon, Zenaida Dove, Cayman Brac Parrot, Black-whiskered Vireo, Vitelline Warbler, Red-legged Thrush and migrant thrushes on passage, and warblers in winter and passage.

9. Bight Road Dry Forest trail

The trail goes from the north to the south coast across the entire bluff, along the western boundary of the Parrot Reserve; accessed from steps on the north coast or on a boardwalk from Major Donald's Drive. This is the best place to see landbirds on the Brac together with (10) below. Look for breeding White-crowned Pigeon, Zenaida Dove, Cayman Brac Parrot, Red-legged Thrush, Black-whiskered Vireo (canopy), Thick-billed Vireo (understorey) and Vitelline Warbler. The Bight Road south trail continues to the southern edge of the bluff where the vegetation becomes xerophytic and orchids bloom in late spring.

10. National Trust Parrot Reserve and Nature Trail

PROTECTED. The Parrot Reserve is dry forest on karstic limestone and there are no trails for access. The Reserve is bisected by Major Donald drive and the National Trust car park is one of the best places to observe parrots on the Island at dawn and dusk as they leave or return to their roosting/ breeding sites. Walk around the National Trust's nature trail loop, signposted to the east off the Bight Road South, especially for parrots, Thick-billed Vireo, Vitelline Warbler, many migrant warblers and thrushes on passage.

11. Eastern Bluff

Park at the Lighthouse at the extreme eastern end of Major Donald's Drive and walk left along the Lighthouse rock trail close to the bluff edge. The Brown Booby colony is around the bluff edge and on ledges and caves below on the bluff face. It is an excellent place to watch the boobies returning from fishing trips at dusk when they are often attacked by Magnificent Frigatebird waiting to steal their fish catch. Peregrine Falcons patrol the cliffs in winter preying on adult and young boobies. Migrating swallows in passage are regular over the bluff shrubland.

12. Lighthouse Walk

This long route is not for the faint-hearted in summer: suitable footwear and a water supply are essential. Access is up Peter's Steps from Spot Bay on the north coast, past Peter's Cave left to the Lighthouse Walk and enter through a gate. White-tailed Tropicbirds nest along the northern bluff face and the trail leads to the Brown Booby colony.

Cayman Brac eastern bluff 55m (150ft) plateau, breeding habitat of Brown Booby on its ledges and caves.

13. The bluff face on the north and south coasts

White-tailed Tropicbirds, in decreasing numbers, nest in holes in the bluff face, as do Barn Owls. Brown Boobies nest on the bluff edge on the south coast west of Ashton Reid Drive.

LITTLE CAYMAN (see map on page 31)

On Little Cayman all the sites are marked on the Department of Tourism's nature tourism brochure.

1. Preston Bay westerly ponds

Two viewing platforms. This wetland on pavement Ironshore has breeding West Indian Whistling-Duck, Green Heron, American Coot and Willet, also wintering herons, duck and shorebirds especially Killdeer.

2. Preston Bay coastal woodland

PROTECTED. Good for migrant vireos and warblers in winter and thrushes in passage, and Summer and Scarlet Tanagers as well as shorebirds, herons and terns on the shore. This area has nesting Sister Islands Rock Iguana, and on the beach, nesting Green and Loggerhead Turtles.

3. Airstrip

Look for Wilson's Snipe (at western end), Peregrine Falcon, Barn Owl, Short-eared Owl, swallows, and Bobolink.

4. Salt Rock Nature Trail

PARTIALLY PROTECTED. Look for breeding Zenaida Dove, Gray Kingbird and Black-whiskered Vireo (in summer), Mangrove Cuckoo and Vitelline Warbler. Also good for wintering and passage migrants including White-eyed and Yellow throated Vireos, many warblers, Dickcissel, Summer Tanager, Indigo Bunting and Rose-breasted Grosbeak.

5. McCoy's Pond

Viewing platform. West Indian Whistling-Duck, rafts of migrant American Coots and Common Gallinules (a few remain to breed under optimum conditions), herons and Yellow Warbler and migrant warblers in the surrounding mangrove.

6. Coastal Mahogany forest

From Spot Bay north on both sides of the road towards Bloody Bay. Migrant warblers, vireos, thrushes and tanagers.

Booby Pond, Little Cayman. A Red-footed Booby colony, together with Magnificent Frigatebird, breeds on the landward side; South Hole Sound on right.

7. Booby Pond Reserve

PROTECTED. Viewing areas and telescopes at the National Trust House (7a). The large Red-footed Booby colony, together with a Magnificent Frigatebird colony, nest in trees inland along the north shore of the pond, where there also is a mixed heronry of Snowy Egrets and Tricolored Herons. West Indian Whistling-Ducks breed here. A major site for all migrant herons, duck, shorebirds, gulls and terns, and raptors. The viewing area on the roof of the Trust House and at the eastern end of the airstrip are excellent sites at dusk to watch returning boobies being chased by waiting frigatebirds attempting to steal their fish catch.

8. South Hole Sound and Owen Island

Look for herons including Reddish Egret, shorebirds including Sanderling, gulls, terns and rare migrants e.g., American Flamingo. Young Red-footed Boobies practice diving in the sound.

9. Kingston Bight

Mixed mangrove and woodland; excellent for unusual migrant landbirds in fall and spring.

10. Tarpon Lake

West Indian Whistling-Ducks breed in the Red Mangrove forest, flocks of Snowy Egrets, Great Egrets and Tricolored Herons forage; also Osprey.

11. Central bluff forest

New roads and development sites are opening up across the island through the dry forest and shrubland. Migrant landbirds.

12. Easterly Ponds

Viewing platforms. Look for flocks of migrating herons, breeding Black-necked Stilts and West Indian Whistling-Ducks. Both Roseate Spoonbill and American Flamingo have wintered here.

13. Coot Pond

Seasonal herbaceous wetland attracts rails including Purple Gallinule and Sora, also Killdeer, Pectoral and Stilt Sandpipers and migrant herons and ducks.

14. East End Viewing area

Doves, migrant warblers and a panoramic view of shrubland with emergent Silver Thatch palms on the eastern bluff. At East Point look across to Cayman Brac and see Brown Boobies and Magnificent Frigatebirds fishing at sea.

15. Grape Tree Ponds

Viewing platform. West Indian Whistling-Duck breed, and migrant ducks and warblers in the mangroves.

16. Jackson's Pond

Viewing platform. West Indian Whistling-Duck breeds. Large flocks of migrant herons occur as well as migrant ducks and rails, preyed on by Merlins and Peregrine Falcons.

INFORMATION FOR VISITING BIRDERS

Planning your trip

There are many international flights to Grand Cayman and Cayman Airways has jet flights to Grand Cayman and Cayman Brac. Cayman Express flies from Grand Cayman to the Sister Islands daily. Birdwatching tours may be booked through the National Trust on Grand Cayman (www. nationaltrust. org.ky) and on the Nature Tourism site on Cayman Brac. The National Trust on Little Cayman provides advice on self-guided tours. The advantages of birding in the Cayman Islands include the ease of accessibility and approachability to the birds. Roadside birding is especially good on Little Cayman and Cayman Brac. Due to the large proportion of migrants in the local avifauna, the optimum time to see the greatest number of species is in fall from late September to November. Spring from late March to mid May is also excellent in most years. There are few forest trails.

Equipment

Off-road birding calls for boots, long trousers, a long-sleeved shirt and a hat as protection against the rough bush as well as the sun. Pack a plastic bag to protect equipment in the rainy months. Always carry water and, if without a guide, a map, compass, GPS and mobile phone.

Electricity

110V and 60 cycles. Carry an adaptor for UK and European plug fittings.

Water and food

Tap water is safe and there are many restaurants on Grand Cayman and Cayman Brac serving excellent local food. On Little Cayman there is one restaurant (and several hotels). Carrying a picnic is good policy for a day out in the bush.

Transport

If not going on a guided tour it is necessary to hire a car and most international brands are available and can be booked online. Or hire a bicycle. Remember to drive on the LEFT!

Security and local customs

Always lock your car and hide valuable items. That being said, crime against tourists remains very low in the Cayman Islands. People are friendly and helpful but farmers are sensitive about persons straying onto private land without permission.

Hazards

There are no poisonous snakes. There are mosquitoes from May to December and sandflies at dawn and dusk outside the winter months so it is always advisable to carry a repellent. Ticks are a problem in pastures with grazing livestock in late winter and spring when the grass is dry. Trousers should be tucked into socks to avoid being infested by small ticks. Spray insect repellent on boots, socks and around the waist. If attacked, remove and wash clothing and remove ticks with tweezers or cover the area with an antihistamine cream. Take care when standing still – if the ground is loosened and grainy it may be a biting ants' nest. On the limestone bluff, there are several vines called 'cow itch' that cause irritation and swelling, so it is inadvisable to grasp plants or tree trunks. The Manchineel tree is poisonous and should be avoided, not even standing under its canopy during rain showers. The jagged-leaved Maiden Plum can cause extreme skin irritation and blistering. Staying on paths will avoid most of these hazards. Although the Islands are small, leaving known trails is not advisable as it is easy to get lost, the terrain underfoot is exceptionally rugged and the vegetation is densely tangled and nearly impenetrable. If travelling alone, tell someone when you intend to return.

HOW TO USE THE FIELD GUIDE

This is the first photographic guide to Cayman Islands' birds, an avifauna that is Neotropical with a large migrant component from the Nearctic region. All species are described, either as regularly occurring species in the main text or as vagrants and rarities in Appendix 6.

NOMENCLATURE

The classification, nomenclature and systematic arrangement follow those of the 7th edition (1998) of the *Check-list of North American Birds* published by the American Ornithologists' Union (AOU) and 52 Supplements (1999–2011). Trinomials follow del Hoyo *et al.* (1992–2012) and Clements (2009). The use of local common names for birds is declining and these are only included if current; all species of heron are called 'gaulin'.

PHOTOGRAPHS

There is at least one image for each species in the main text. There are several illustrations for endemic subspecies to show less known (and photographed) species from different angles, and to include dimorphic plumages, and immature and winter plumages for migrants where possible. Yves-Jacques Rey-Millet took the great majority of the photographs in the wild in the Cayman Islands or, in a limited number of cases where this was not possible, in Cuba, the Dominican Republic, Costa Rica and North America. Additional photographers are listed at the front of the book. The month in which the photographs were taken was captioned and, for breeding species, the Island(s) on which breeding occurs. Photographs of the habitats and aerial views of the Islands are included in the Introduction. The Grand Cayman Thrush (almost certainly extinct) is illustrated in Appendix 1.

SPECIES ACCOUNTS

Taxonomy

Species are classed as monotypic (lacking subspecies) or polytypic (having two or more recognised subspecies). The number of subspecies worldwide is shown in parentheses for the latter.

Description

L refers to the body length, from the tip of the bill to the tip of the tail, and it is given in centimetres (inches in parentheses). Differences between males and females (sexual dimorphism) are noted where relevant. Measurements were taken from Raffaele *et al.* (1998). Descriptions of the morphology of males, females, immatures and/or juveniles are included based on the author's field and museum observations. Non-breeding plumage is usually described first because most migrants occur predominately in this plumage in the islands. Breeding and juvenile/immature individuals are also described for most species. Shorebirds are regularly seen in early fall in full, partial breeding or juvenile plumage and in full or partial breeding plumage in late spring.

When observing a new bird, first note the general size and shape. For waterbirds note bill and leg size and colour. Note plumage: in ducks note colour of head and back and the speculum in flight; in plovers note breast-bands; in sandpipers notes bill length, rump and tail patterns and wing-stripe, in gulls and terns note colours of tips of wings and tail and tail shape in terns. For landbirds look for bill shape and presence or lack of an eye-ring and wing-bars, and in flycatchers note shape of crown and colour of lower mandible. Warblers present a challenge in fall with 30 species in various plumages, especially the dull first-winter and intermediate non-breeding adult plumages. Note the most obvious colours, followed by presence or absence of wing-bars, eye-stripes, supercilia, tail spots (white on the outer rectrices), streaking on mantle or on underparts, as well as rump and leg colour. The quickest identifications are made using multiple field characteristics combined with identification of voice, the habitat and any distinctive habits, such as tail bobbing or the probing feeding technique of shorebirds.

Similar Species

Dominant field characters are described that usefully distinguish the differences between similar species and that under discussion.

Voice
Vocalisations of the great majority of birds are translated into words to mimic sounds, based on field observations and recordings, including the work of G. B Reynards, Cornell Laboratory of Ornithology, New York and Alexandra Guenther-Colhoun. The songs of resident birds are diagnostic; the great majority of migrants do not sing but most make short diagnostic calls that help with identification. 'Silent' in the text refers to those migrants that usually do not sing or call while in the islands.

Habitat and behaviour
The preferred habitats (dominant forest trees are described in the **Vegetation and habitats** section in the Introduction and Appendix 2) indicate the most likely locations of nesting or foraging birds, both resident and migrant. Behavioural traits contribute to locating and identifying species. These include voice, choice and methods of prey capture, tail flicking, head bobbing and understanding seasonal changes, either in habitat use (e.g., parrots are cavity nesters in forest from April-July and are less visible during these months) or in yearly distribution patterns (e.g. the majority of migrants occur between August and May and are absent in summer). For ease of reading, vegetation/habitats are given in lower case in the text.

   Accounts of breeding behaviour may include breeding months, courtship, nest shape, clutch size and average brood size. 'Fledglings' refers to newly hatched young in the nest or still covered in down and unable to fly, and 'juvenile' refers to fledged birds in their first plumage feathers and able to fly. Most landbirds nest in spring and early summer, from March to July. Some have a longer season, from February to September, raising more than one brood. A few, e.g., the Bananaquit, breed throughout the year. Large seabirds, such as sulids, tropicbirds and frigatebirds, raise only one young during a lengthy nurturing process. Two species of migrant tern breed in early summer. West Indian Whistling-Duck and mixed colonies of herons breed at different times on the three Islands in different years, depending on the rains and food availability. Additional breeding data is welcomed; please send to pebrad@candw.ky.

Range
The majority of species breed only in the Americas (including the West Indies); some species also breed in other zoogeographic regions (Palearctic, Cosmopolitan and Pantropical) defined on page 45. The islands in the Caribbean Sea include the West Indies biogeographical region (Bahamas, Greater Antilles, Cayman Islands, Providencia, Swan Islands and Lesser Antilles), as well as the insular Caribbean islands of Netherland Antilles (Aruba, Bonaire and Curacao), Trinidad and Tobago, and islands and cays off the coast of Venezuela and Middle America. Range distribution maps are not included in the text as the majority of birds are dispersed in habitats throughout the three Islands; for breeding birds confined to one or two Islands see Appendices 3 and 4. For migrant birds, the wintering non-breeding range in the Americas and the Bahamas, Greater Antilles and Cayman Islands is given, as well as the migration routes on passage; the Lesser Antilles is sometimes included if relevant. Restricted range species (confined to a small area of the Caribbean region) and West Indian endemic species are mentioned here (Bradley *et. al.* 2006).

Status
Describes the breeding, migrant, resident or introduced status of each species and its distribution, frequency and abundance in preferred habitats. Each of the three Islands is mentioned where variations occur. Species of global conservation concern, from the IUCN Red list (2009), are mentioned here. There are three categories of Threat: 'Critically Endangered', 'Endangered' or 'Vulnerable', with a fourth category, 'Near-Threatened', for species approaching Threatened status. In the Cayman Islands there are three existing species of Global Conservation Concern (Table 1). The Grand Cayman Thrush is now considered extinct; otherwise it would be classified as critically endangered. Species deemed to be of regional concern, not listed by IUCN, such as breeding colonial seabirds are also listed (Bradley & Norton 2009).

Table 1. Global conservation status of three bird species.

Vulnerable: Species facing a high risk of extinction in the wild	Near-Threatened: Species not presently in the 'threatened' categories, but close to qualifying, or likely to qualify, in the near future
West Indian Whistling-Duck *Dendrocygna arborea*	Cuban Parrot *Amazona leucocephala* Vitelline Warbler *Setophaga vitellina*

Terms of abundance

Abundant	Observed in flocks or in colonies in preferred habitat
Common	Regular in numbers in preferred habitat
Fairly common	Likely to be seen in preferred habitat
Uncommon	Unlikely to be seen even in preferred habitat
Local	Seen irregularly in part of the preferred habit
Rare	Very unlikely to be seen even in preferred habitat
Introduced	Feral populations introduced by man, breeding in the wild
Casual	Visitor that may occur in any season. Individuals may be long-term, non-breeding stayers
Irregular	Occurs in winter or on passage but not every year
Vagrant	Fewer than ten records

Abbreviations

CMW	Central Mangrove Wetland, Grand Cayman
IB	Intermittent breeder, does not breed every year
MB	Migrant breeder (summer)
RB	Resident breeder
?B	Breeding suspected, but not proven
FB	Former breeder, no recent breeding records

Zoogeographic regions and range definitions

Middle America	Refers to Mexico and the Central American countries of Honduras, Nicaragua, Belize, Costa Rica, Guatemala and El Salvador.

The New World

Nearctic	North America
Neotropics	Middle and South America and the Caribbean
Holarctic	Palearctic and Nearctic regions combined

The Old World

Palearctic	Europe, north Africa and Asia north of the Himalayas
Pantropical	within the tropics, refers to seabirds in this text
Cosmopolitan	occurs in most of the zoogeographic regions

BIRD TOPOGRAPHY

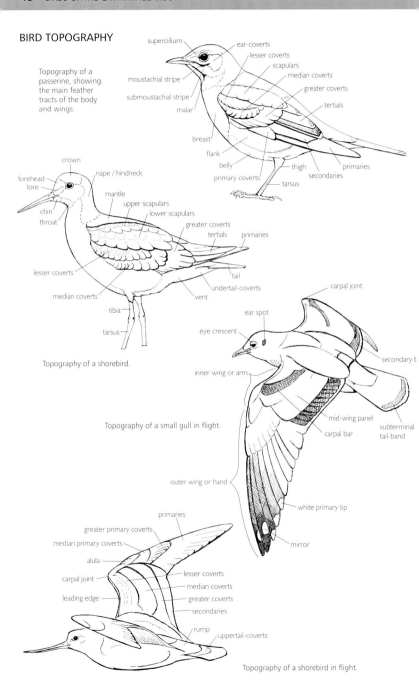

Topography of a passerine, showing the main feather tracts of the body and wings.

superciilium
ear-coverts
lesser coverts
scapulars
median coverts
greater coverts
tertials
moustachial stripe
submoustachial stripe
malar
breast
flank
belly
primary coverts
thigh
tarsus
secondaries
primaries

crown
nape / hindneck
forehead
lore
mantle
upper scapulars
lower scapulars
chin
throat
greater coverts
tertials
primaries
lesser coverts
tail
undertail-coverts
median coverts
vent
tibia
tarsus

Topography of a shorebird.

carpal joint
ear spot
eye crescent
secondary b
inner wing or arm
mid-wing panel
carpal bar
subterminal tail-band

Topography of a small gull in flight.

outer wing or hand
white primary tip
mirror

primaries
greater primary coverts
median primary coverts
alula
carpal joint
leading edge
lesser coverts
median coverts
greater coverts
secondaries
rump
uppertail-coverts

Topography of a shorebird in flight.

Black-bellied Whistling-Duck
Dendrocygna autumnalis

Taxonomy Polytypic (2).
Description L 46–53cm
(18–21in). Goose-like. Adult
has pale eye-ring, grey face and
upper neck; cinnamon lower
neck, breast and back; black
abdomen, rump and tail; long
pink legs and orange bill. Large
buff and white patch on wing
shows on upperwing-coverts in
flight. Immature is paler with
grey legs.
Similar species West Indian
Whistling-Duck has black and
cream pattern on flanks, black
bill and legs, upperwing-coverts
are silver-grey in flight.
Voice Silent.
Habitat and behaviour
Herbaceous wetlands and
lagoons. Grazes on pondweed
and grasses. Head and legs are
carried low in flight.
Range *D. a. autumnalis* breeds
in southern Arizona, Texas,
Middle America to Panama. In

Adult. Orange bill and legs and black abdomen are diagnostic, May.

the West Indies it is casual in
Cuba, vagrant elsewhere in the
Antilles. *D. a. discolor* breeds in
South America.

Status Casual visitor or vagrant,
single adults on Grand Cayman,
and one resident 1994–1996.

Adult. Grey face with prominent white eye-ring, May.

West Indian Whistling-Duck
Dendrocygna arborea

Adult.

Local name Whistler.
Taxonomy Monotypic.
Description L 48–56cm (19–22in). Long-legged, long-necked, upright duck, similar to a goose. Adult has chestnut-brown forecrown, crested blackish hindcrown and nape, dark brown back, wing-coverts edged buff, black flanks with large creamy-white central patch on each dark feather. Face grey-buff, lower neck and breast rich tawny chestnut with black streaks, whitish abdomen and undertail-coverts spotted with black, rump and tail blackish; bill, legs and feet greyish-black. In flight head and tail are held low, legs project beyond tail, silvery inner primaries and greater primary coverts and unmarked underwings. Juvenile smaller, duller with less spotting.

Adult. Note dark flank feathers with creamy central areas, February.

Similar species Black-bellied Whistling-Duck has black abdomen, red legs and feet, and broad white patch on upperwing-coverts in flight. Fulvous Whistling-Duck smaller with cinnamon underparts, black tail with white subterminal band, in flight dark upperwing is unmarked.

Voice Far-carrying 3–5 syllable whistle *tsssee-tsssee-tssee-seee (rising)-tsweer.*

Habitat and behaviour Vegetarian, taking fruits (Royal Palm), bulbs, grass seeds. Frequently up-ends for weed with small invertebrates attached, on large saline lagoons, Herbaceous wetlands and temporarily flooded grasslands; also grazes on beach vegetation (Ambrosia and Bay Vine) on shores of marine sounds, Little Cayman. Breeds in all months, peaks April–September on Grand

Adult has silvery inner primaries and greater primary coverts and legs trail in flight, July.

Cayman, and January–May on Little Cayman and Cayman Brac. Territorial and aggressive. Nests above ground or in roots of Red Mangrove forest and Buttonwood shrubland, in dry forest and in tall grassland near water. Clutch up to 13 white eggs, young in creches of up to 30 supervised by several adults. Close affinity with man, adopting gardens with ponds, especially where fed.

Range West Indian endemic species; its range has contracted with significant populations remaining only in the Bahamas, Turks and Caicos Islands, Cuba, Dominican Republic, Cayman Islands and Antigua and Barbuda. It is uncommon in Jamaica, Puerto Rico and Martinique.

Status Common resident on Grand Cayman and Little Cayman and has a small population on Cayman Brac. Retreat into dense cover during breeding and moult periods. The only duck to breed in the Islands, its recovery is due to conservation efforts and a feeding station; some illegal hunting remains but the greatest threat is now habitat loss and feral dogs. Global Conservation Status: Vulnerable.

Flock at Preston Bay wetlands, Little Cayman, April.

Fulvous Whistling-Duck
Dendrocygna bicolor

Adult. Note barring on back and pale elongated flank feathers.

Taxonomy Monotypic.
Description L 46–51 cm (18–20in). Large goose-like duck. Adult has face, breast and underparts unmarked cinnamon, dark brown back barred tawny, defined buffy-white elongated flank feathers. In flight head held low and legs extend beyond tail, shows blackish unmarked upperwing, cinnamon underparts, white uppertail coverts contrast with dark rump and tail.
Similar species West Indian Whistling-Duck is dark brown with black and cream markings along flanks; in flight shows a silvery patch on upperwing. Black-bellied Whistling-Duck has black abdomen, white eye-ring, red legs and feet; in flight shows large white patch on upperwing-coverts.

Voice Two-syllable whistle *wu-cheu*.
Habitat and behaviour Vegetarian, forages in similar habitats to West Indian Whistling-Duck.
Range Breeds in North America from southern California and United States Gulf coast to southern Florida, both coasts of Middle America to South America, and the West Indies: where abundant in Cuba, locally common in Hispaniola, and rare in the Bahamas, Jamaica and Lesser Antilles.
Status Rare passage migrant and casual long-staying visitor.

Adult.

Gadwall
Anas strepera

Adult female. Only duck with white secondaries (speculum), November.

Taxonomy Monotypic.
Description L 46–57cm (18–22in). Dabbling duck similar to Mallard. Adult non-breeding male and female, and immature, have round head, thin bill with grey upper mandible and yellow sides, plumage speckled and mottled brown; in flight has white underwing-coverts and all plumages show diagnostic white inner secondaries often visible at rest. Male breeding has brown head, grey overall with speckled breast, black rump and tail, long silver tertials and cinnamon-edged scapulars.
Similar species Mallard has violet-blue speculum; male has yellow bill and female has dark patch in centre of yellow bill. Female American Wigeon has blue bill, green speculum and white on forewing.
Voice Silent.
Habitat and behaviour Herbaceous wetlands, singly with Blue-winged Teal and American Wigeon.
Range Holarctic. Breeds in North America, where its range is expanding; winters to the Gulf coast and Florida, Mexico, and rarely in the West Indies: in the Bahamas, Turks and Caicos Islands and western Cuba; vagrant elsewhere in the Antilles.
Status Vagrant or rare in winter and passage.

Adult male. Note the scalloping on the breast.

American Wigeon
Anas americana

Adult male with white crown and flank patch, January.

Local name Baldpate.
Taxonomy Monotypic.
Description L 46–56cm
(18–22in). Large dabbling duck
with rounded head, dark patch
around eye, small pale blue bill
with black border and tip, and
pointed tail. Eclipse male and
adult female have grey speckled
head and tan underparts; adult
female has dark ring around
base of bill. Breeding male
has whitish crown and wide
iridescent green streak from eye
to nape, pinkish-brown breast,
long scapulars, white abdomen
and flank patch. In flight both
sexes show green speculum;
male has white upperwing-
coverts, grey in female.
Similar species Blue-winged
Teal and Green-winged Teal are
smaller with dark bill, females
are speckled overall; in flight
Blue-winged Teal has blue
upperwing-coverts, grey in
male Green-winged Teal.

Voice Silent.
Habitat and behaviour Saline
lagoons and herbaceous
wetlands, usually with mixed
flocks of ducks. Feeds close to
the water's edge and cover,
seldom in open water.
Range Breeds in northern and
central North America; winters
from western North America
across central and southern

United States and Middle
America to northern Colombia,
and the West Indies, common
in Cuba and Hispaniola,
uncommon in the Bahamas,
rest of Greater Antilles and
Cayman Islands; rare in the
Lesser Antilles.
Status Uncommon winter
visitor and passage migrant,
October–May.

Adult female, December.

Blue-winged Teal
Anas discors

Adult male, February.

Local name Teal.
Taxonomy Monotypic.
Description L 38–40cm
(15–16in). Small dabbling duck
with long blackish bill, yellow
legs and feet. Adult female
and eclipse male speckled grey-
brown edged buffy, black eye-
line; adult female has broken
white eye-ring, white throat
and pale spot at base of bill. Fall
males in eclipse until November;
face crescent emerges from
October. Breeding male has
wide white crescent curving
backwards from front of eye,
dark head, black and buff
spotted plumage, white flank
patch, black undertail-coverts.
In flight both sexes shows light
blue upperwing-coverts, male
has green speculum, brown
in female. Juvenile similar to
female.
Similar species American
Wigeon female has blue bill
and white or grey upperwing-
coverts. Green-winged Teal
female has black eye-line, lacks
white spot on lores; both sexes
show grey upperwing-coverts in
flight. Northern Shoveler female

has large spatulate bill.
Voice Females quack loudly in
flocks.
Habitat and behaviour All
wetland habitats, dabbling
for weed and insects: in large
flocks on saline lagoons and in
small flocks and individual pairs
on most inland ponds. Flies up
vertically on take-off.
Range Breeds in North America
and locally on the Gulf coast

and Florida; winters from the
southern United States to
South America and the West
Indies; most common migrant
duck in Bahamas, Greater
Antilles and Cayman Islands.
Status Common winter visitor
and very common passage
migrant, July–June. Numbers
fluctuate greatly from year to
year depending on wetland
conditions.

Adult female with dark eye-line and whitish spot at base of bill, February.

First-winter male showing blue upperwing-coverts and green speculum, December.

Northern Shoveler
Anas clypeata

Adult male in breeding plumage.

Taxonomy Monotypic.
Description L 43–53cm (17–21in). Large dabbling duck with very wide, spatulate bill. Adult female and eclipse male (to October) have greyish-brown plumage broadly edged with buff; male bill is greenish-black, female has dark yellowish-black bill edged orange (can appear bright yellow in the field). Breeding male has iridescent blackish-green head and neck, black back streaked with white, yellow iris, white breast, chestnut flanks with large white patch before black stern, orange legs and black bill. First-winter male resembles adult female, may have white on face, black bill and green showing on head by February and chestnut on sides emerging by April. In flight adult male has light blue upperwing-coverts and green speculum, extensive white on underwing-coverts, and spatulate bill emphasises size of head.

Similar species Only duck with large spatulate bill. In flight Blue-winged Teal smaller, adult male shows crescent on face. American Wigeon male has white upperwing-coverts, grey in female. Green-winged Teal has dark upperwing-coverts.
Voice Usually silent.
Habitat and behaviour Saline lagoons and herbaceous ponds with other ducks.
Range Holarctic breeder. Breeds in northern and central North America; winters from the United States to northern South America and the West Indies: common in Cuba, uncommon to rare in the Bahamas, rest of Greater Antilles and Cayman Islands; rare and local in the Lesser Antilles.
Status Uncommon to fairly common passage migrant and winter visitor, August–April.

Adult female. Bill is dark on top with orange edges, iris gold, November.

First-winter male, black bill, golden eye and green developing on head, December.

Northern Pintail
Anas acuta

Adult male has blue-grey sides to bill and white stripe from nape to breast, May.

Taxonomy Polytypic (3).
Description Male 72cm (28in), female 55cm (21in). Slender dabbling duck with rounded head, long neck and wings and pointed tail. Adult female is mottled light brown with dark bill and reddish edges to feathers on upperparts. Eclipse male has greyish plumage, pale brownish head and white neck. Breeding male has dark brown head, nape and throat; white stripe from nape to white breast, bright blue-grey sides to bill, silver-grey flanks, white flank patch, black tail with long central tail feathers; both sexes have long scapulars. In flight both show dark upper-wings with white trailing edges; male has green speculum.
Similar species Green-winged Teal and Blue-winged Teal females are smaller with short tail and, with Northern Shoveler, show more white on underwings in flight. American Wigeon female has green speculum. In all four species white is absent on trailing edge of wing.
Voice Silent.
Habitat and behaviour Freshwater wetlands and saline lagoons with other ducks.
Range Holarctic breeder. *A. a. acuta* breeds in northern North America; winters from North America and Middle America to Panama and northern South America, Bermuda, and the West Indies: common in Cuba, uncommon to rare in the Bahamas, Hispaniola and Puerto Rico and the Cayman Islands.
Status Vagrant or rare winter visitor and passage migrant, October–April.

Adult female with rounded head and pointed tail, August.

Green-winged Teal
Anas crecca

Adult male with chestnut head and iridescent green crescent onto nape, February.

Taxonomy Polytypic (3).
Description L 33–39cm (13–15.5in). Small brown dabbling duck. Eclipse male and adult female mottled greyish-brown with buffy edges to feathers, pale buff on undertail-coverts, small bill. Breeding male has chestnut head and neck, iridescent green band from eye to nape, greyish back with long scapulars, buffy breast spotted cinnamon, flanks finely barred grey with vertical white bar on breast side, pale abdomen, black bill. In flight both sexes show dark upperwing-coverts separated from green speculum by white or gold bar; male has white underwings and abdomen contrasting with grey sides and flanks. **Similar species** Blue-winged Teal female has white lores and both sexes show blue upperwing-coverts in flight. Northern Pintail larger with longer neck, shows less white on underwing, and white trailing edge to inner wing.
Voice Silent.
Habitat and behaviour Usually solitary. Flooded marl pits, brackish and herbaceous wetlands.
Range Northern hemisphere. *A. c. carolinensis* breeds in northern North America; winters throughout North America to southern Florida, rarely to Middle America, and to the West Indies where is fairly common in Cuba and uncommon to rare in the rest of Greater Antilles and Cayman Islands.
Status Rare winter visitor and passage migrant, October–March.

Adult female has green speculum with inner brown bar on upperwing, December.

Ring-necked Duck
Aythya collaris

Adult male has white edges to blue bill and white subterminal band, June.

Taxonomy Monotypic.
Description L 40–46cm (16–18in). Diving duck with domed crown. Breeding female is brownish overall with dark brown crown, back and breast, white eye-ring and post-ocular line, whitish patch at base of dark grey bill, which is black-tipped with thin white subterminal band. Eclipse male resembles breeding male but duller, indistinct white ring at base of bill, brownish-grey flanks. Breeding male has iridescent violet-black head and neck, greenish-black back; blue bill with white edges, white subterminal ring and wide black tip, narrow white band at base of bill, black breast with white vertical bar in front of wing, flanks finely barred pale grey, rest of underparts white. In flight both sexes show dark upperwing-coverts, grey secondaries and pale underwings.
Similar species Lesser Scaup

has no white on bill, back is greyish (appears white in field); female lacks white eye-ring and post-ocular line. In flight both show white secondaries with dark trailing edge to wing.
Voice Silent.
Habitat and behaviour In large flocks on saline lagoons, singles or small flocks on herbaceous wetlands, and flooded marl pits and artificial ponds on Grand Cayman.

Range Breeds in northern North America; winters in the United States and Middle America to Panama and the West Indies, where common in the northern Bahamas and Cuba, uncommon in Jamaica, and regular in the Cayman Islands; rare or vagrant elsewhere.
Status Uncommon to locally common winter visitor and fall passage migrant, October–February.

Adult female. White eye-ring and post-ocular line are diagnostic, March.

Lesser Scaup
Aythya affinis

Adult male has blue bill with small black tip, September.

Taxonomy Monotypic.

Description L 38–46cm (15–18in). Diving duck with rounded head, yellow iris, peaked hind-crown and blue bill with black terminal spot. Breeding female has wide white patch at base of bill; dark brown head, breast and back, paler on flanks. Eclipse male has brownish wash overall, no white at base of bill. Breeding male appears black and white in the field; iridescent black head and neck, breast and tail, back speckled grey, flanks speckled pale greyish-white. In flight both show white secondaries with dark trailing edge. First-winter resembles adult female.

Similar species Ring-necked Duck male has white subterminal band around blue bill, white bar in front of wing and black back; female has white eye-ring and post-ocular line. In flight both show pale grey secondaries.

Voice Silent.

Habitat and behaviour In large flocks on saline lagoons, and small groups on brackish ponds in Buttonwood and herbaceous wetlands.

Range Breeds in northern North America; winters from North America to northern South America and the West Indies where fairly common in northern Bahamas, Cuba, Jamaica and Cayman Islands; uncommon elsewhere in the Greater Antilles.

Status Fairly common to locally common winter visitor, October–April.

Adult female has oval white patch at base of bill, February.

Red-breasted Merganser
Mergus serrator

Adult male has a white collar, secondaries and coverts.

Taxonomy Monotypic.
Description L 51–64cm (20–25in). Diving duck. Adults have shaggy crest and thin red bill hooked at tip. Breeding male has iridescent green head, wide white neck ring, back black with extensive white on wing-coverts, breast rufous streaked with black, and grey flanks. Breeding female has brownish rufous crest, whitish chin, back and sides barred grey and buff. Eclipse male similar to female. In flight body looks elongated; male shows dark unmarked head, white neck ring and white upperwing-coverts and secondaries; female has white secondaries and dark upperwing-coverts.
Similar species Hooded Merganser (vagrant) is smaller with dark bill and longer tail; female has rounded crest and brown back, sides and flanks.
Voice Silent.
Habitat and behaviour Marine sounds, open sea near coasts and saline lagoons.
Range Holarctic breeder. Breeds in northern North America; winters in North America including the Gulf coast to Florida, and very rarely in the West Indies: in the northern Bahamas, Cuba, Dominican Republic, Puerto Rico and Cayman Islands.
Status Rare winter visitor and passage migrant, November–February.

Adult female has a white speculum.

Ruddy Duck
Oxyura jamaicensis

Male has acquired non-breeding plumage by August.

Taxonomy Polytypic (3).
Description L 35–43cm
(14–17in). Small, compact,
stiff-tailed duck. Adult female
and first-winter have dark
brown upperparts, buffy-white
cheek crossed by black bar, bill
grey. Breeding male is reddish-
brown overall with black head,
white face and wide blue bill;
eclipse male has head and
upperparts greyish-brown,
bill blue-grey. In flight shows
unmarked dark upperwing and
long tail.
Similar species None.
Voice Silent.
Habitat and behaviour
Solitary on artificial lakes, saline
lagoons.
Range *O. j. jamaicensis* breeds
from North America to Middle
America, with sedentary
populations in the West Indies
in the Bahamas, Caicos Islands
and Greater Antilles. North
American birds winter south
in the breeding range. Two
subspecies breed in South
America.
Status Rare passage migrant,
April–June.

Adult female and juvenile have a brown bar across cheek, August.

Pied-billed Grebe
Podilymbus podiceps

Adult breeding plumage has black throat and subterminal band on bill, April.

Local names Diver, Diving Dapper.
Taxonomy Polytypic (3).
Description L 30–38cm (12–15in). Small stocky diving bird, brownish-grey plumage, large flattened head with blackish crown, pale face, white eye-ring and fluffy white undertail-coverts. Breeding adult has black chin, throat and thick whitish conical bill with black subterminal band. Non-breeding have pale throat, bill with partial ring and darker plumage. Young juvenile greyish with black and white striped face, becoming brownish-grey with faint ring on bill as immatures.
Similar species Least Grebe (vagrant) is blackish with yellow iris and small pointed bill.
Voice Highly vocal with long loud, harsh *ca-cac-cao-wo-kwop-kwop*…over 20 syllables and descending, and pairs have a similar but faster duet; a single low *kuk;* wailing and short barks; chicks and juveniles *peep* constantly.

Habitat and behaviour On all wetlands, preferring freshwater and brackish ponds, and flooded marl pits on Grand Cayman; rafts of migrants

Adult non-breeding has brownish plumage, faint partial ring on bill and pale chin, March.

Legs are set far back to allow efficient diving and running on water's surface for take-off.

Breeds throughout the year, 3–8 whitish eggs, on raft of plant matter, either built up from the pond bottom or floating and anchored to adjacent vegetation at the water's edge. Young with red facial skin are often carried on parents' backs.

Range *P. p. antillarum*, West Indian subspecies, is resident, sedentary and breeding in the Bahamas, Greater Antilles and Cayman Islands, becoming rare on smaller islands. *P. p. podiceps* breeds from North America to northern South America; winters in southern breeding range including the West Indies.

Status Fairly common breeding resident on Grand Cayman and Cayman Brac; uncommon sporadic breeder on Little Cayman. The population, mainly on Grand Cayman, is swelled by rafts of migrants of *P. p. podiceps* in winter and on passage, October–May.

are regular on saline lagoons. Floats high in the water, but responds to threat by sinking with only head showing or diving to surface far off. Takes aquatic insects and small fish.

Young have striped heads with orange spot on crown; note reddish bill and eye-ring, October.

American Flamingo
Phoenicopterus ruber

Adults are long-legged and long-necked with black-tipped triangular bill.

Taxonomy Monotypic.
Description L 108–122cm
(42–48in). Unmistakable. Adult
is very large, entirely coral-pink
with very long neck and legs,
and very heavy decurved bill:
pale at base with large black
tip. In flight shows black
primaries and secondaries.
Immature smaller with greyish
plumage and black-tipped grey
bill: pink plumage develops
over three years.
Similar species None.
Voice Usually silent, may *honk*
on take-off.
Habitat and behaviour
Filter-feeds on saline lagoons
and marine sounds for algae
and very small invertebrates
pumping water through
inverted bill.
Range The Caribbean,
breeding on the Yucatan coast,
Bonaire, on Inagua in southern
Bahamas, Caicos Islands,
eastern Cuba, and Hispaniola;
wanders widely in the region.
Status Casual visitor in all
months. Single adults, more
frequently immatures, may stay
for several months.

Adults in a marine sound on Cayman Brac, April.

White-tailed Tropicbird
Phaethon lepturus

Adult, May.

Local name Boatswain Bird.
Taxonomy Polytypic (5).
Description L 81cm (32in) includes 30–40cm tail streamers. Adult resembles a tern in flight except for diagnostic long central tail streamers; brilliant white plumage, black streak through eye, orange decurved bill, black on outer primaries and band across inner upperwing-coverts. Juvenile has upperparts heavily barred black and white, yellowish bill, and lacks tail streamers from pointed tail. In adult plumage by third year.
Similar species None.
Voice Constant *cri-et cri-et-cri-et* and *crit-crit crit* heard over long distances, adult screeches and chick screams and hisses in the nest hole when disturbed.
Habitat and behaviour Pelagic, only coming ashore to breed.

Adult in flight, May.

Beautiful acrobatic displays, with frequent calling, in inshore waters before flying directly into nest holes. Plunge-dives for squid and flying fish, chased and robbed by frigatebirds and preyed on by wintering Peregrine Falcons. Nest with one egg in crevices and holes in the bluff face from late January: entrance may be hidden behind overhanging vegetation. Breeding is prolonged, from laying to fledging is c.18 weeks, between January–July, otherwise pelagic.
Range Pantropical. Western Atlantic subspecies *P. l. catesbyi* breeds in Bermuda, the Bahamas, Turks and Caicos Islands and Greater Antilles: where it is common, and on Cayman Islands, Virgin Islands and Lesser Antilles: where it is uncommon.
Status Summer breeding migrant. Small colony dispersed

Adults have black on outer primaries and across coverts, May.

along the south coast bluff from Pedro to Beach Bay, Grand Cayman and a larger colony around the coastal bluff of Cayman Brac, December–September. Numbers have declined sharply on Cayman Brac and declines continue throughout its range due to habitat loss and predation; considered threatened in the region.

Adults displaying close to nest on southern bluff, May.

Magnificent Frigatebird
Fregata magnificens

Local name Man O' War.
Taxonomy Monotypic.
Description L 94–104cm (37–41in). Very large black seabird. Long narrow, pointed wings bent at the wrist give shallow M silhouette, long deeply forked tail, long bill hooked at tip, small red feet. Male is entirely glossy purple-black, except when red gular sac inflated during courtship and early breeding season. Female larger, brownish-black except for white breast and diagonal bronze bars across upperwing-coverts. Juvenile (1–2 years) is brownish black with white head and throat; immature (2–4 years) has whiter head and underparts.
Similar species None.
Voice Bill clattering at nest.
Habitat and behaviour Breeds Little Cayman, September–April, when males fly with gular sacs inflated. Stick nests built in mangrove and dry shrubland and dry forest, 1.5–4m elevation, with adults constantly stealing nesting material dislodging eggs and

Breeding male with gular sac partially deflated, November.

young of boobies and other frigatebirds. Both parents brood one large white egg for 45–55 days and feed the white fluffy fledgling, which fledges after 5–6 months. Most adult males leave the colony, April–September. Feeds on squid and flying fish snatched from the surface, also by

Breeding male with gular sac fully inflated, November.

kleptoparasitism robbing both booby species and tropicbirds of fish. Roosts on cays in North Sound, Grand Cayman and the eastern bluff, Cayman Brac. **Range** Breeds on Atlantic and Pacific coasts of South America, Gulf coast of Texas, Yucatan Peninsula and Belize, Marquesas Key (Florida), and locally in the West Indies: with

Adult female has white breast, January.

the largest colonies in Barbuda, Cuba, Tobago, Puerto Rico and Virgin Islands, and smaller colonies in the Bahamas, Hispaniola, Cayman Islands, Anguilla, Redonda and St Lucia and the Cape Verde Islands off West Africa.
Status Abundant breeding resident, colony of 400–1,000 birds breeds in the centre of the Red-footed Booby colony on the landward side of Booby Pond, Little Cayman. Non-breeding resident on Grand Cayman (former breeder up to 1950s), regular in George Town and Morgan's Harbours, and Cayman Brac.

Juvenile in late second year, Little Cayman only, June.

Female on nest sheltering single fledgling on edge of Booby Pond, Little Cayman only, February.

Masked Booby
Sula dactylatra

Adult is distinguished from white phase Red-footed Booby by black mask and tail.

Taxonomy Polytypic (4).
Description L 81–91cm (32–36in). Adult is white with black mask around yellowish bill, yellow iris, black greater coverts and tail; legs and feet olive-yellow. In flight shows black primaries and secondaries. Juvenile has dull dark brown head and upperparts, white collar and underwings. Adult plumage gradually develops by third year.
Similar species White phase Red-footed Booby lacks black mask and has white tail and red feet.
Voice Male hisses and clicks bill on nest.
Habitat and behaviour Eastern bluffs of Cayman Brac; forages at sea together with flocks of Brown Boobies, plunge-dives for squid and flying fish. An adult paired with a Brown Booby was observed on many occasions on the bluff: either no eggs were produced or the single egg was infertile.
Range Pantropical. *S. d. dactylatra* breeds very locally on cays off the coast of the Yucatan Peninsula with the largest colonies in the West Indies on Pedro Cays, Jamaica and Puerto Rico; also small colonies on the Virgin Islands, Anguilla and Redonda.
Status Vagrant or rare visitor on Cayman Brac, where two birds remained over 20 years.

Adult has black greater coverts and flight feathers on upperwing.

Brown Booby
Sula leucogaster

Adult female, Cayman Brac only, April.

Local name Booby.
Taxonomy Polytypic (4).
Description L 71–76cm
(28–30in). Adult has dark
chocolate upperparts, head
and neck to mid-breast, sharply
defined from the white lower
breast, abdomen, undertail-
and underwing-coverts; yellow
legs and feet. Male has yellow-
grey bill with pink and blue
facial skin at base; female bill is
pinkish–yellow with black spot
in front of the eye. Juvenile is
entirely dull brown, paler on
the abdomen, with greyish bill,
legs and feet. Immature has
abdomen mottled with white.
Adult plumage after two years.
Similar species Juvenile Red-
footed Booby is greyish-brown
overall with white tail and
rump, dull orange feet.

Voice Female grunts and honks
and male whistles and croaks
on the nest.
Habitat and behaviour Roosts
and breeds on the bluff edge,
on ledges and caves in the
bluff face, Cayman Brac. Adults
forage in flocks at sea, flocks

Adult showing white abdomen and underwing, Cayman Brac only, April.

Immature has abdomen mottled brown and white for up to 2 years, Cayman Brac only, April.

Juvenile, Cayman Brac only, April.

scrape with a few strands of vegetation on the eastern bluff, on the southern bluff edge: where nests are among xeric shrubland with Century Plants; a few individual nests on the rocky shore since 2006. Eggs incubated up to 47 days, but only one young is ever raised, as siblicide (killing) by the larger occurs. Latter fledges after 90 days and is dependent for up to nine months.

Range Pantropical. Caribbean subspecies *S. l. leucogaster* breeds on cays off the coast of the Yucatan Peninsula, Panama, Colombia and Venezuela, and throughout the West Indies: with the largest colonies in Puerto Rico, Anguilla, Virgin Islands, the Bahamas and Hispaniola.

Status Common breeding resident on Cayman Brac. Vagrant to Grand Cayman; mainly immature birds after storms.

often following fishing boats, or singly in inshore waters; plunge-dive for fish and squid. Pursued by frigatebirds for fish and adults and young preyed on by overwintering Peregrine Falcons. Breeding throughout much of the year, with peaks November–January and May–June. Following prolonged courtship, 1–2 white eggs laid on the ground in a

Developing juvenile plumage, 12–14 weeks, Cayman Brac only, April.

Red-footed Booby
Sula sula

Taxonomy Polytypic (3).
Description L 66–76cm
(26–30in). Smallest booby, with
pointed head and long wings.
Occurs in two colour morphs.
All adults have white rump,
tail and undertail-coverts;
bright red legs and feet, and
bluish-grey bill with pink facial
skin. Golden wash develops on
the head and neck in breeding
plumage. Ten percent of the
colony are white morphs,
with brilliant white plumage
except for black primaries and
secondaries: shows dark carpal
patches below in flight. Ninety
percent are brown morphs,
entirely soft brown except for
white tails, rumps and lower
abdomens and dark brown
primaries and secondaries.
Occasional intermediates occur.
Juvenile is entirely greyish-
brown, paler below, dark
breast-band and dull greyish
yellow feet; rump and tail are
greyish-brown. Adult plumage
by end of second year.

Brown morph with red feet and white tail, Little Cayman only, January.

Adult brown morph with white rump and tail, March.

Similar species Adult Masked
Booby has black mask and
black tail; juvenile Brown
Booby is dark brown overall
with olive-yellow feet.
Voice Rattled *gra gra gra
gra*, squawks at nest, and bill
clacking between pairs.
Habitat and behaviour Birds
forage long distances for fish
and squid, returning at dusk
when they are harried by
waiting frigatebirds. Breeding
season October–June, with
prolonged courtship. Stick
nests built in mangrove and
dry forest and shrubland on
the landward side of Booby

Adult white morph showing black flight feathers and carpal patch on underwing.

White morph on nest, Little Cayman only. Note bright bill and facial skin, February.

Pond; one white egg. Young raised by both parents, fledging after 110 days and fed for a further six months.
Range Pantropical. Caribbean subspecies *S. s. sula* breeds on cays in the southern Caribbean, Belize and the West Indies with the largest colonies in the Cayman Islands, Hispaniola (Navassa), Grenada and the Grenadines, Puerto Rico, and smaller colonies in the Virgin Islands, Redonda and Tobago; casual in the Bahamas and Jamaica.
Status Abundant resident, colony on Booby Pond, Little Cayman. Estimated at 4,800 pairs in 1997 census; recent population declines are partly due to effects of hurricanes.

Vagrant to Grand Cayman and Cayman Brac. The colony is one of the largest in the Caribbean, protected by National Trust Law and as a Ramsar Site (wetland of international importance). Birds are thought to disperse to other colonies in the region but return to the natal colony to breed.

Brown morph on nest, Little Cayman only. Note bright bill and facial skin, December.

Double-crested Cormorant
Phalacrocorax auritus

Adult shows orange throat pouch and green iris, January.

Taxonomy Polytypic (5).
Description L 74–89cm (29–35in). Large, erect, iridescent black seabird, wings blackish-brown edged black, long neck, long thick bill hooked at tip, orange lores and throat pouch (seldom visible), green iris and stiff tail. Breeding adult has two ear tufts (not easily visible). First-winter is brownish with pale buffy neck and breast, dark abdomen, yellow throat pouch and lower mandible.
Similar species Anhinga has silvery coverts and very long tail.
Voice Silent.
Habitat and behaviour Buoyant flight with neck kinked. Single birds usual after storms perched on mangrove at edge of lagoons, marine sounds, and freshwater ponds on Grand Cayman. Swims with only head exposed and dives to pursue fish; adopts a 'wing-spread' position with wings held out to dry.
Range *P. a. floridanus* breeds in the United States from North Carolina to Florida and the Gulf Coast, and in the West Indies in the northern Bahamas and Cuba. Fairly sedentary although northern birds disperse south to the Bahamas and Cuba; casual elsewhere in the northern West Indies. Three additional subspecies breed in North America, and *P. a. heuretus* breeds in the Bahamas and Cuba.
Status Uncommon migrant and long-stay casual visitor. In all months with some remaining for extended periods.

First-year. Note yellow throat pouch and lower mandible and buffy breast, December.

Anhinga
Anhinga anhinga

Taxonomy Polytypic (2).
Description L 85cm (34in).
Unmistakable. Adult has long
sinuous neck, long fan-shaped
white-tipped tail, thin pointed
yellow bill. Male is entirely
black apart from silvery white
scapulars and upperwing-
coverts. Female wings similar
but head, neck and breast
buffy-pink. Juvenile and
immature smaller and browner
than female, with reduced
white on back and wings. In
flight soars and glides on long
pointed wings showing black
underwings and silvery-white
on upperwings.
Similar species Double-crested
Cormorant lacks silvery coverts,
has shorter tail.
Voice Silent.
Habitat and behaviour The
Ironshore, perched in mangrove
on marine sounds and saline
lagoons. Swims with head
and upper neck above water
and dives to pursue fish. Holds
wings extended to dry after
swimming and usually flies with
neck outstretched.
Range *A. a. leucogaster* is
mainly sedentary and breeds
in south-eastern United States
including Florida through the
Gulf coast of Mexico, Middle

Adult female, November.

America to Panama, and the
West Indies in Cuba; vagrant
elsewhere. Another subspecies
breeds in South America.
Status Rare passage migrant
in any season, and occasional
long-stay visitor.

Adult male, December.

Brown Pelican
Pelecanus occidentalis

Taxonomy Polytypic (6).
Description L 107–137cm (42–54in). Very large, heavy seabird with extremely large, long bill and expandable pouch. Non-breeding adult has pale yellow head, white neck, dark bill and iris, grey upperparts. First-winter has brownish-grey plumage, whitish underparts and brown iris. Breeding adult (seldom observed) has yellow crown, white hind crown extending as a stripe down foreneck, chestnut hindneck, pale grey coverts and back, dark underparts, white iris.
Similar species None.
Voice Silent.
Habitat and behaviour Marine sounds, harbours, roosting on saline lagoons. Flies with slow, deep wing strokes and

Non-breeding adult has white head and nape, December.

First-winter has brownish head and nape and pale abdomen, December.

plunge-dives for fish. Small flocks mainly of juveniles and immatures, most probably in post-breeding dispersal, arrive in late fall and remain until early spring; practice diving in shallows along Seven Mile Beach.
Range *P. o. occidentalis* breeds in the West Indies. *P. o. carolinensis* breeds United States coast of South Carolina, on cays off the Atlantic-Gulf-Caribbean and south to northern South America.
Status Uncommon to locally fairly common winter visitor, passage migrant and casual long-stay visitor in all months; both subspecies probably occur.

Least Bittern
Ixobrychus exilis

Adult male stalking, Grand Cayman only, March.

Taxonomy Polytypic (5).
Description L 28–35cm
(11–14in). Smallest heron. Adult
buffy with dark upperparts, long
neck, chestnut nape and wings
with bright buffy-yellow patch
on upperwing-coverts, two
white lines on back, throat and
breast striped white and buff,
abdomen and undertail-coverts
white, legs and feet yellowish,
bill yellow with dark on upper
mandible. Male has crown and
back greenish-black, chocolate
in female. Juvenile like female
but back greyish and greyish-
buff patch on coverts.
Similar species Green Heron is
larger, lacks buffy coverts, neck
and breast are rufous.
Voice Low rapid three-syllable
hu hu hoo descending, harsh
rik-rik-rik, and warning call *kek*.
Habitat and behaviour Widely,
though sparsely, distributed
in herbaceous wetlands with
surrounding sedge and cattails;
also around saline lagoons.
Usually solitary, secretive,
'freezes' with neck pointing
skywards; takes fish, aquatic
insects, molluscs, toads and
lizards by stalking or holding
unto cattails and lunging.
Breeds, April–July, in nest
of cattails and sedge, 2–3
bluish-white eggs, incubated
by female.
Range *I. e. exilis* breeds from
eastern North America to
Middle America and the West
Indies, where it is common in
Cuba, Jamaica and Puerto Rico.
North American birds winter
from the Greater Antilles to
northern South America.
Status Uncommon breeding
resident on Grand Cayman, and
uncommon winter visitor and
passage migrant, August–May.
Numbers in decline due to loss
of freshwater habitat.

Adult female has chocolate-brown upperparts, January.

Great Blue Heron
Ardea herodias

Taxonomy Polytypic (5).
Description L 107–132cm (42–52in). Very large heavy-bodied heron with long thick neck and sturdy yellowish bill with dark upper mandible. Adult is greyish overall with white head and throat, wide black supercilium forming trailing plumes, pale buffy-grey neck with central black and white streaking to breast, black 'shoulders', tops of legs chestnut, long legs greyish. In flight shows black flight feathers and grey coverts. Juvenile/immature similar but grey overall with entirely dark crown and streaked neck.
Similar species Only large grey heron.
Voice Usually silent, except when alarmed *braak*.
Habitat and behaviour
All wetland habitats: saline lagoons, marine sounds, coastal fringing reefs, the Ironshore and herbaceous ponds. Solitary except on migration and at mixed heron roosts on mangrove cays. Walks slowly or stands motionless to hunt large fish, rodents, kittens and reptiles. Flight slow with deep wingbeats and neck folded.
Range *A. h. occidentalis* breeds from Florida to islands off

Adult non-breeding, April.

Venezuela, and West Indies where it is uncommon on Cuba, St Thomas and St Croix, and Anegada, mainly sedentary populations. *A. h. herodias* breeds in North and Middle America, and *A. h. wardi* breeds in United States and winters south to northern South America and West Indies, where common in the Greater Antilles:

both subspecies probably occur in the Cayman Islands.
Status Uncommon winter visitor and fairly common passage migrant, mainly August–May, although present in all months.

First-year has dark crown, streaked neck and brown on coverts, June.

Adult showing black flight feathers, November.

Great Egret
Ardea alba

Taxonomy Polytypic (4).
Description L 89–107cm (35–42in). Large, slender white heron with yellow bill, blackish legs and feet. Juvenile similar but smaller and thinner. Breeding adult develops long feathery dorsal plumes.
Similar species All other white herons are smaller; Snowy Egret adult has black bill and yellow feet. Immature Little Blue Heron has blue-grey bill and greenish legs and feet. White phase Reddish Egret has bicoloured bill, and juvenile has black bill.
Voice Usually silent, except for alarmed *krro-aar* on take-off.
Habitat and behaviour All wetland habitats including flooded grassland, reef-protected marine shores, and rough pasture including roadsides (similar habitat to Cattle Egret) and, increasingly, in urban areas. Stalks prey and takes fish, insects, snakes, lizards and small birds. Solitary, except on migration, at feeding aggregations or in mixed heron roosts.
Range Cosmopolitan. *A. a. egretta* breeds throughout the Americas and the West Indies in Jamaica and Cuba and, uncommonly, in the Bahamas. Northern birds winter south through the West Indies.

Only adult heron with yellow bill and black legs, March.

Status Common winter visitor and locally very common passage migrant, July–early June; a few over-summer.

Adult stalking and displaying dorsal breeding plumes, March.

Adult non-breeding, September.

Snowy Egret
Egretta thula

Breeding adult displays crest and dorsal plumes, diagnostic yellow feet and fine black bill, March.

Taxonomy Polytypic (2).
Description L 51–71cm (20–28in). Medium-sized heron. Adult is entirely white with needle-like black bill, yellow lores and feet, and black legs. Breeding adult develops white crest, breast plumes and aigrettes on back. Juvenile smaller with black fore-leg and yellow-green stripe on hind leg, greenish-yellow feet, yellowish bill with black tip; immature resembles adult with black bill but retains yellow leg stripe.
Similar species Immature Little Blue Heron has heavier blue-grey bill, entirely greenish legs and feet. White phase Reddish Egret is larger with bicoloured bill, and juvenile has black legs and feet. Immature Great Egret is much larger with yellow bill. Cattle Egret has yellow bill.
Voice Harsh *gaarrh* and low *kroo* in feeding aggregations.
Habitat and behaviour Core species in mixed heron feeding aggregations on lagoons and ponds, and also singly on the Ironshore, fringing reefs, flooded grassland and herbaceous wetlands. Active forager, flying over ponds in flocks with the leading group 'foot-dragging' yellow feet in flight to attract prey; also stalks prey, waits motionless on a branch, and shuffles feet in mud; takes mainly fish; also crabs, aquatic

Snowy Egret immature showing yellow stripe on hind-leg and yellow feet, January.

insects, grasshoppers, snakes, toads and snails. Breeds in mixed heronries (always with Tricolored Heron), builds stick nest platforms, 2–4 bluish eggs, December–May, in forest in the Central Mangrove Wetland, Grand Cayman and in dry forest behind Booby Pond and Jackson's Pond, Little Cayman. Young fledge after 30 days.

Range *E. t. thula* breeds from North America to South America, and the West Indies in the Bahamas, Greater Antilles, Cayman Islands, Virgin Islands, and Lesser Antilles. North American birds winter in southern breeding range.

Status Abundant breeding resident on Grand Cayman and Little Cayman, and common to abundant winter visitor and passage migrant, August–May. Numbers reduced in summer (and early dry seasons) on Cayman Brac and Little Cayman following post-breeding dispersal and migration.

Juvenile has yellowish-olive feet and stripe on hind-leg, January.

Little Blue Heron
Egretta caerulea

Taxonomy Monotypic.
Description L 56–71cm (22–28in). Medium-sized slender heron. Adult has purple head and neck, dark slate-blue body, bluish-grey bill with black tip slightly downturned, bluish lores and olive legs. Juvenile is white with grey primaries, grey lores, chalky blue bill, and greenish legs; when changing to adult plumage becomes mottled grey-blue and white.
Similar species White phase Reddish Egret adult is larger with bill pink proximally; dark phase adult has rufous head and neck; juvenile has black bill, grey legs and feet (bluish in breeding adult). Immature Great Egret is larger with yellow bill and black legs. Cattle Egret is smaller with yellow bill. Immature Snowy Egret has black bill, yellow lores and front of legs black with yellowish-green stripe on hind-leg.

Adult, March.

First-summer immature transitioning to dark adult plumage, May.

Juvenile is entirely white with bill blue-grey at base and greenish-olive legs, Grand Cayman, February.

Voice Silent.

Habitat and behaviour Often solitary on the Ironshore, fringing reefs, reef-protected shores and flooded grassland, and at the edge of mixed heron feeding aggregations on lagoons. Takes mainly fish, also crabs, aquatic insects, toads and snails. Flies with neck usually folded, extended only for short flights. Congregates at same winter roosts as Snowy Egret and breeds intermittently in Snowy Egret and Tricolored Heron colonies on Grand Cayman and Little Cayman, December–May, in a stick platform nest.

Range Breeds in southern United States to central South America, and the West Indies. North American birds winter throughout the breeding range.

Status Fairly common resident, winter visitor and passage migrant, mainly August–May, and a rare intermittent breeder on Grand Cayman and Little Cayman. The reduced summer population is of immature (white) birds.

Tricolored Heron
Egretta tricolor

Taxonomy Polytypic (2).
Description L 61–71cm (24–28in). Medium-sized slender heron with long thin neck, yellow lores, long black-tipped bill, and yellowish legs. Adult is blue-grey, with white stripe on central neck to grey breast, white abdomen and underwing-coverts. In peak breeding plumage (seen briefly) adult has bright blue lores and bill, red iris, red legs, nuchal plumes, long pale chestnut back plumes and shaggy grey neck feathers. Juvenile has bright rufous neck, mantle and wing-coverts gradually becoming mottled with grey.
Similar species Adult Little Blue Heron and dark phase Reddish Egret have no white on underparts.
Voice Harsh *gaarh* similar to Snowy Egret.
Habitat and behaviour In large feeding aggregations (mainly with Snowy Egret) on saline lagoons, and singly in all wetland habitats, including marine sounds, fringing reefs and herbaceous wetlands taking mainly fish, also crabs, insects, toads and snails. In large mixed heron roosts in winter. Often flies with neck

Adult non-breeding has yellow facial skin, February.

extended. Breeds in mixed heronries with Snowy Egret in mangrove forest, Grand Cayman, and dry forest, Little Cayman and Cayman Brac, December–May. Stick platform nest, 2–3 bluish eggs, and young fledge after 35 days.
Range *E. t. ruficollis* breeds from the Atlantic/Gulf coasts of North America to northern South America and the West Indies, in the Bahamas, Greater Antilles, Cayman Islands and, rarely, in the Lesser Antilles. North American birds winter south through the breeding range.
Status Abundant breeding resident on Grand Cayman and Little Cayman and uncommon on Cayman Brac. Common to abundant winter visitor and passage migrant, September–May. Numbers decline in summer on Cayman Brac and Little Cayman following post-breeding dispersal and migration.

Juvenile has rufous on neck and wing-coverts, March.

Reddish Egret
Egretta rufescens

Adult dark morph non-breeding has bill less brightly coloured, February.

Taxonomy Polytypic (2).
Description 69–81cm (27–32in). Medium-large, thickset heron occurring in white and dark colour morphs. Adult of both morphs has pale iris, long, deep bill is diagnostic among herons with distal half black and proximal half dull pink (non-breeding) and bright pink with blue facial skin (breeding). Immature and juveniles of both morphs have bill black or blackish, dark grey lores, legs and feet. Dark morph adult is grey with shaggy reddish head, neck and breast; juvenile is chalk grey with pale rufous head and neck, pale iris and dark lores, bill and leg. White morph adult is entirely white with dark legs. Both morphs have shaggy plumes on neck when breeding.
Similar species Tricolored Heron adult has white on

Immature dark morph with all-dark bill, June.

Juvenile/first year, dark morph moulting, March.

underparts; juvenile smaller with brighter rufous neck and coverts. Immature Great Egret has yellow bill and black legs. Snowy Egret is slender with thin black bill and yellow feet.
Voice Silent.
Habitat and behaviour Usually solitary on saline lagoons and marine sounds. Foraging behaviour is diagnostic, dashing, jumping and turning with wings extended, also forms wing canopy shading water to attract prey.
Range Central America and northern South America. *E. r. rufescens* breeds in southern United States in Florida and coastal Texas, and in areas along the Caribbean coast of Middle America (Yucatan, Belize, Costa Rica) and Colombia, and the West Indies in the Bahamas, Cuba, and Hispaniola and Puerto Rico (rare). It is mainly sedentary with some wandering and post-breeding dispersal, and winters in the breeding

Breeding adult bill is pink proximally and black distally, March.

range to Venezuela.
Status Uncommon casual visitor in all months. Occurs singly, white and dark morphs equally frequent. Some remain for extended periods.

White morph immature with black bill, exhibiting diagnostic foraging behaviour, May.

Cattle Egret
Bubulcus ibis

Taxonomy Polytypic (3).
Description L 48–64cm
(19–25in). Small, stocky,
short-necked white heron
with short, thick yellow bill,
yellowish legs and dark feet.
Breeding adult has red iris, bill
and legs (shown briefly at nest)
and buffy-orange crown, neck,
breast and dorsal patches;
similar but paler feathers and
dark yellowish legs and feet
often shown on non-breeding
adults. Juvenile is white with
dark bill and legs.
Similar species All are larger.
Snowy Egret has black bill and
yellow feet. Juvenile Little Blue
Heron has blue-grey bill and
olive legs.
Voice *Breeck* and croaks on
the nest.
Habitat and behaviour
Terrestrial, in rough and wet
pastures, including airports
and urban areas. Stalks insects,
lizards, mice and follows
livestock that flush prey. Breeds
in mixed heronries, December–
May, in mangrove forest, Grand
Cayman, and dry forest, Little
Cayman and Cayman Brac.

Adult breeding develops orange plumes on crown, neck and back, Little Cayman, April.

Stick nest platform with 2–3
greenish eggs.
Range Cosmopolitan. *B. i.
ibis* breeds in the Americas,
and the West Indies (since
1944). Populations from the
eastern United States migrate
to the Greater Antilles in large
post-breeding dispersals. This
subspecies also breeds in Africa,
India and Europe.
Status Uncommon breeding
resident and fairly common
to locally abundant winter
visitor and passage migrant. It
colonised *c.*1957.

Adult non-breeding, October.

Juvenile has dark bill and legs.

Green Heron
Butorides virescens

Local name Mary Perk.
Taxonomy Polytypic (4).
Description L 40–48cm
(16–19in). Small, short-legged
and short-necked heron; long
bill with dark grey upper and
yellowish lower mandible;
yellow lores, legs and feet. Adult
has crested steel-blue head and
back, dark petrol-blue wings
with coverts edged whitish,
rufous neck and breast, white
band from throat to breast, grey
abdomen and flanks, and short
dark tail. Breeding plumage
shows brighter rufous, with grey
dorsal plumes and orange legs.
Juvenile has dark upperparts,
spotted wing-coverts, rufous
and white streaks on necks and
breast.
Similar species Least Bittern
is smaller with cinnamon-buff
wing-covert patch.
Voice Loud s*k-yew* when
flushed; warning *kek-kek-kek*.
Habitat and behaviour Highly
territorial and usually solitary,
on all wetlands including the
Ironshore, fringing reefs, wet
pasture, and among coconut
palms in urban/littoral areas.

Adult has steel-blue head and back, wing-coverts edged cream.

Flies with neck outstretched
or retracted, often hunched
when perched. Stalks fish,
aquatic insects, shrimp, crabs
and lizards or perches and
waits motionless. Rough stick
platform nest, 2–3 greenish-
blue eggs, from 0.3–3m up at
the edge of lagoons and ponds,
and close to sink holes on the
bluff. Breeds throughout the
year on Grand Cayman.
Range *B. v. virescens* breeds
from the United States, Middle
America to Panama, and

throughout the West Indies;
winters from the southern
United States to northern South
America. *B. v. bahamensis* is
confined to the Bahamas and is
sedentary.
Status Fairly common breeding
resident on Grand Cayman
where breeds throughout
the year, uncommon to fairly
common on Little Cayman and
Cayman Brac. Migrants occur in
winter, August–May. Decline in
western Grand Cayman is due
to loss of habitat.

Adult foraging in herbaceous wetland,
May.

Juvenile has heavily streaked neck and breast and white spots on coverts, October.

Black-crowned Night-Heron
Nycticorax nycticorax

Taxonomy Polytypic (4).

Description L 58–71cm (23–28in). Medium-sized, stocky, short-necked heron with heavy, sharply pointed black bill with paler base to lower mandible and short yellow legs. Adult has black crown, nape and back; grey coverts without pale edging; white face, neck and throat, red iris and pale grey underparts. Breeding adult has nuchal plumes, red lores and legs. Juvenile is brownish with large spots on wing-coverts and upperparts, wide cream streaks on buffy underparts, bill with yellowish lower mandible. In flight legs usually do not extend fully beyond tail.

Similar species Juvenile Yellow-crowned Night-Heron is greyer, bill thicker, blunt and curved towards tip, smaller spots on wing-coverts, greater wing-coverts are edged pale (diagnostic), longer legs with feet extending fully beyond tail in flight.

Voice *Quork*.

Habitat and behaviour Usually

Adult, March. Bill varies from all blackish to having yellowish on straight lower mandible and base of upper mandible.

solitary, but flocks in migration. Crepuscular, nocturnal and

Juvenile has spots on wing-coverts and creamy streaks on face and underparts, February.

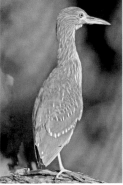
Juvenile lacks pale edges to greater coverts, February.

prefers foraging in herbaceous wetlands and brackish ponds, flooded marl pits and wet pastures, Grand Cayman; roosts in mangroves. Lack of freshwater habitat is the most likely reason for its rarity on the Sister Islands.

Range. Cosmopolitan. *N. n. hoactli* breeds from North America to central South America, and the West Indies in the Bahamas and Greater Antilles; winters throughout the West Indies.

Status Fairly common winter visitor and locally common passage migrant on Grand Cayman, August–May, where a few, mainly immatures, over-summer. Rare on passage on Cayman Brac and Little Cayman.

Yellow-crowned Night-Heron
Nyctanassa violacea

Local name Crabcatcher.
Taxonomy Polytypic (6).
Description L 56–71cm
(22–28in). Medium-sized heron;
large, black head with white
crown and cheeks giving a
striped pattern to face, dark
thick rounded bill, red iris and
yellow legs. Neck, mantle and
underparts are pale grey, dark
grey wing-coverts edged with
pale grey. Breeding adult has
nuchal plumes and yellow wash
on white crown. Juvenile has
greyish upperparts, thin brown
and greyish-cream streaks on
underparts, wing-coverts finely
spotted, greater wing-coverts
edged buff, bill blackish.
Intermediate immature lacks
juvenile spots on coverts and
face pattern of adult. Legs
extend fully beyond tail in flight.
Similar species Juvenile
Black-crowned Night-Heron is
browner, stockier, has pointed
bill with yellowish lower

Adult has dark grey back and wing-coverts edged with pale grey, March.

Second-year transitioning to adult plumage. Spots on coverts absent, wing-coverts have pale edges, December.

Juvenile. Pale edges to greater wing-coverts are diagnostic for this species, February.

mandible, wing-coverts have larger spots and greater coverts are not edged pale, feet do not extend fully beyond the tail in flight.

Voice *kowrk* and *qwok*, louder and higher pitched than Black-crowned Night-Heron.

Habitat and behaviour In all habitats where this nocturnal crustacean specialist can take land, hermit, mangrove and ghost crabs. Breeds in mixed heronries, January–May, in mangrove forest and shrubland, Grand Cayman and Little Cayman, and small single-species colonies or single pairs in dry forest. Bulky stick nest with 2–3 greenish-blue eggs.

Range Americas. *N. v. bancrofti* breeds coastal regions from Baja California to El Salvador and the West Indies. North American birds winter to South America.

Status Fairly common breeding resident, formerly very common throughout the three Islands with the greatest decline on Grand Cayman; fairly common in winter and locally common on passage, August–May.

Adult and juvenile have heavy blunt bill curved towards tip on upper and lower mandible unlike Black-crowned Night-Heron which has straight edge to lower mandible.

White Ibis
Eudocimus albus

Taxonomy Monotypic.
Description L 56–71cm (22–26in). Entirely white wading bird with blue iris, red facial skin, legs and feet, long decurved orange bill. Four distal primaries and tail show black in flight. Breeding adult has deeper red facial skin, gular sac and legs, bill black distally. Juvenile has dark brown back and wings, neck mottled brownish-white, white underparts, pinkish-grey bill and legs; primaries and secondaries show black in flight. White adult plumage emerges from first spring.
Similar species No white herons have long decurved red bill or black tips to primaries. Immature Glossy Ibis is dark overall.
Voice Silent.
Habitat and behaviour Herbaceous wetlands preferred, also lagoons, the Ironshore and flooded grassland. Forages by walking slowly in shallow water for fish, crustaceans, molluscs and insect larvae and on land probes for insects, spiders and

Adult, February.

lizards. Flies with neck and legs outstretched.
Range Mainly sedentary, breeds from southern United States to both coasts of South America to Peru and French Guinea and the West Indies in Cuba, Jamaica and Hispaniola;

uncommon in winter in the Bahamas and Puerto Rico.
Status Increasingly uncommon, long- and short-stay casual visitor in all months, Grand Cayman, mainly August–May; rare on passage, Little Cayman and Cayman Brac.

Juvenile, October.

Juvenile transitioning to adult plumage, April.

Glossy Ibis
Plegadis falcinellus

Non-breeding adult. Head and neck are finely streaked, March.

Taxonomy Monotypic.
Description L 56–64cm (22–25in). Appears black in the field and in flight. Non-breeding adult has head and neck finely streaked whitish, iris brown, dark facial skin with white at base of bill, iridescent green-bronze on back and wings, long grey decurved bill and dark olive legs. Juvenile entirely dark. Breeding adult has chestnut head, neck and body.
Similar species Juvenile White Ibis has white underparts.
Voice Silent.
Habitat and behaviour Herbaceous wetlands and flooded grassland preferred, also saline lagoons and the Ironshore; roosts in fringing mangrove and Buttonwood in eastern Grand Cayman. Forages in shallow water by walking slowly for fish, crustaceans, molluscs and insect larvae, and on land probes for insects, spiders and lizards. Flies with neck and legs extended.
Range Cosmopolitan. Breeds in eastern North America from Atlantic/Gulf (locally) to Costa Rica and Venezuela, and the West Indies in the Greater Antilles; resident and non-breeding in the Bahamas. Northern birds winter south of the breeding range.
Status Locally common to uncommon passage migrant and casual visitor in all months, mainly August–June.

Adult in breeding plumage.

Roseate Spoonbill
Platalea ajaja

Adult (right) and pale juvenile (left) at nesting colony, Florida.

Taxonomy Monotypic.
Description L 66–81cm (26–32in).Very large, pink and white wading bird with very long, flattened, grey bill ending in spoon-like tip. Greenish skin on bald head, white neck with black band around nape, whitish mantle, red wing-coverts on pink wings, uppertail ochre with pink rump, dark pink clump of feathers on breast, reddish legs. Colours brighter in the breeding season. Juvenile has white feathered head, neck, back and breast; rest of plumage greyish with pink wash, tail pink, dark legs.
Similar species None.
Voice Silent.
Habitat and behaviour Saline lagoons. Walks slowly sweeping bill in a wide arc in shallow water taking small fish, insects and weed. In flight appears entirely pink with neck outstretched and slow wingbeat.
Range Breeds in the United States from the Gulf coast of Texas and southern Florida through Middle America to South America, and the West Indies in the Bahamas in Great Inagua and Andros, Cuba and Hispaniola; casual or vagrant elsewhere.
Status Rare casual visitor or vagrant, usually immature birds, occasionally remaining for extended periods.

Adult flies with neck outstretched and spoon visible.

Turkey Vulture
Cathartes aura

Adult eating fish on shore, May.

Local Name John Crow
Taxonomy Polytypic (4).
Description L 68–80cm
(27–32in). Large raptor with
broad wings and small head.
Adult is dark brown, red bill
pale distally and hooked,
head and neck have bare skin
varying from pinkish to red,
and reddish feet. Juvenile is
dull brown with grey skin on
the head and neck. In flight
shows silvery wing-coverts and
flight feathers, tail long.
Similar species None.
Voice Silent.
Habitat and behaviour Large
wing area allows soaring and
tilting flight using thermals.
Forages using highly developed
sense of smell; almost
exclusively a scavenger, feeding
on carrion of rats, mice, snakes
and birds, often found as road

kills and on shores.
Range *C. a. aura* breeds in
western North America to
Coast Rica, northern Bahamas
and Greater Antilles. North
American birds winter through
the southern breeding range
to South America.
Status Non-breeding, casual
visitor, observed in all months
on Grand Cayman and Cayman
Brac; rare on Little Cayman.
Some become resident for
several years.

Adult. Note silvery flight feathers and
long tail.

Osprey
Pandion haliaetus

Local Name Fish Hawk
Taxonomy Polytypic (4).
Description L 53–61cm (21–24in). Sexes alike, female larger and heavier. *P. h. carolinensis* has dark brown upperparts and wide band from eye to hindneck, white head with dark streaks on crown, white neck and underparts with streaks on breast varying in density depending on sex (darker on females sometimes forming a dark necklace). The Caribbean *P. h. ridgwayi* has very pale head and faint mark on ear coverts. In flight, wings are long and narrow angled at the carpal joints into a shallow V, white underparts and underwing-coverts contrast with black carpal patches and barred primaries, secondaries and tail. Immature has mantle and wing-coverts edged pale.
Similar species None.
Voice Far carrying *s-ee-ew* repeated, alarm *kui kui kui*, and series of whistles.
Habitat and behaviour Fringing reefs, marine sounds and all wetland lagoons and ponds. Makes spectacular plunge dives, with wings folded

Adult subspecies *carolinensis*, non-breeding migrant, November.

and long legs extended; also takes fish near surface without a dive. Diet mainly fish but, increasingly, egrets are taken at the three Island's airports and ducks and shorebirds from lagoons.
Range Cosmopolitan. *P. h. carolinensis* breeds in North America and winters to South America and the West Indies though usually south of the range of sedentary populations of the paler Caribbean *P. h. ridgwayi* which breeds in Cuba, the southern Bahamas and Belize.
Status Uncommon winter visitor and fairly common to locally common passage migrant in small flocks; mainly August–May, but observed in all months. *P. h. ridgwayi* is very occasionally resident for long periods.

Adult showing barred flight feathers and black carpal patches on under-wing, October.

Adult holding fish lengthways in talons, October.

Swallow-tailed Kite
Elanoides forficatus

Taxonomy Polytypic (2).
Description L 51–66cm (20–26in). Unmistakable black and white slender silhouette in flight. Shining white plumage apart from black primaries and secondaries on long pointed wings and very long, black, deeply forked tail. Immature is similar with pale wash on underparts and tail less forked.
Similar species None.
Voice Silent.
Habitat and behaviour Majority seen in migration in leisurely flight over the Islands, very occasionally perched in mangrove forest.
Range Central and South America. *E. f. forficatus* breeds from coastal south-eastern United States to South America. North American birds winter in South America and migrate through the northern Bahamas and the western Greater Antilles.
Status Regular passage migrant

Adult on migration, July.

in fall, July–October, with peaks in September, and uncommon from March–May. Numbers have increased with groups over 40 now seen on Grand Cayman; fewer observations on Cayman Brac and Little Cayman.

Adult showing blue-grey and black on upperparts, July.

Northern Harrier
Circus cyaneus

Other name Hen Harrier
Taxonomy Polytypic (2).
Description L 46–61cm
(18–24in). Slender hawk. Adult
has owl-like face and small
bill, distinctive white uppertail
coverts, and long narrow
barred tail. Male has blue-grey
head, upperparts, upperwings
and breast, rest of underparts
whitish; in flight shows whitish
underwings, black tips to
primaries and black trailing
edge to wing. Female is larger,
with dark brown upperparts,
wings and tail, and heavy rusty
streaking on buffy underparts;
in flight underwings are heavily
barred. Juvenile similar to
female but darker above and
cinnamon below.
Similar species None.
Voice Silent.
Habitat and behaviour
Flies low with tilting glides
over mangrove shrubland
on the edge of lagoons, wet
meadows and herbaceous
wetlands hunting for birds,
especially young gallinules,
stilts and coots. Also in littoral
areas or perched on the
ground or a low bush hunting

Adult male has grey upperparts, black tips to primaries and diagnostic white rump.

for mice, lizards, toads, crabs,
insects.
Range Holarctic. *C. c.
hudsonius* breeds in North
America and northern Mexico;
winters south to Middle
America, Panama and northern
Colombia, irregular in the
Greater Antilles except Cuba
where it is common.
Status Irregular; in different
years ranges from rare to fairly
common in winter and on
passage, October–April.

Adult female has brown upperparts, buff underparts heavily streaked and diagnostic white rump.

Adult female flushing rails on lagoon, November.

Red-tailed Hawk
Buteo jamaicensis

Taxonomy Polytypic (14).
Description L 48–64cm (19–25in). Large hawk, in flight adult shows broad wings with dark bar on leading edge and red tail with dark subterminal band. Dark brown mottled upperparts; underparts vary from whitish-buff to reddish with dark abdominal band. Juvenile has underparts streaked, bar on abdomen and brown tail barred.
Similar species No other members of this genus have been observed.
Similar species None.
Habitat and behaviour Soars high on rounded wings. Juveniles occur in mangrove forest during tropical storms.
Range Americas. *B. j. borealis* breeds in eastern North America and Middle America, *B. j. jamaicensis* breeds in the northern Bahamas, Greater Antilles (except Cuba) and northern Lesser Antilles. Mainly sedentary, although northern

Adult, June.

North American birds migrate into the southern breeding range.
Status Rare or vagrant on Grand Cayman in January and March. Many unconfirmed observations at high elevations.

Adult showing dark band across abdomen and red tail, May.

American Kestrel
Falco sparverius

Taxonomy Polytypic (17).
Description L 23–30cm (9–12in). Both adults have slate grey head, rufous crown patch, white face with black malar stripe below eye and second vertical stripe over ear coverts; small, hooked black-tipped bill, rufous back barred black, long blunt wings and long rufous tail. Male has grey upperwing-coverts spotted black, pale cinnamon underparts spotted blackish, unbarred tail with black subterminal band and white tip. Female larger with rufous wing-coverts finely barred black, breast whitish heavily streaked cinnamon, and finely barred tail. Juvenile resembles adult.
Similar species Merlin has no white on face, malar stripe less distinct, and underparts darkly streaked in both sexes.
Voice '*Killy-killy-killy*', heard very infrequently.

Adult male has grey upperwing-coverts.

Habitat and behaviour
Solitary. Forages in all habitats. Perches on telegraph wires and trees to dive for lizards and insects, or hovers with spread tail to detect prey; catches dragonflies on the wing.
Range Breeds from North America to South America, and West Indies where it is mainly sedentary (*F. s. dominicensis* in Hispaniola and *F. s. sparveroides* in Cuba and the Bahamas; both subspecies occur in Jamaica). Northern North American *F. s. sparverius* winters south to Panama including the Cayman Islands.
Status Fairly common winter visitor and passage migrant, November–April.

Adult female has brown upperwing-coverts and heavily streaked breast, February.

Merlin
Falco columbarius

Adult female, *F. c. columbarius* has pale supercilium. Note white spots on dark flank feathers, December.

Taxonomy Polytypic (9).
Description L 24–34cm (10–13.5in). Small, compact falcon with dark crown, one indistinct malar stripe, whitish chin and throat, black tipped bill with yellow cere, yellow feet and legs. Underparts whitish; breast, sides and flanks heavily streaked dark brown or blackish as Cayman migrants show great colour variation Male has plain dark bluish or blackish-grey upperparts, and 2–5 thin pale grey bands on black tail. Female heavier with dark brownish upperparts, tail dark brown with pale buff bands. Juvenile similar to female. Broad wings show sharply pointed in flight.
Similar species American Kestrel has two pronounced black stripes on white face, rufous back and tail, cinnamon underparts not heavily streaked. **Voice** Silent.
Habitat and behaviour Solitary. In all habitats, but more closely associated with wetlands. Fast direct flight with powerful downward stroke, aggressively attacking prey from the air, especially waders on lagoons, white-winged doves and swallows over open land; also bats, and takes insects from perch.
Range Holarctic. *F. c. columbarius* breeds in northern North America, and winters south of the breeding range to northern South America from Ecuador to Brazil, and the West Indies.
Status Fairly common winter visitor and passage migrant, July–May.

Adult male has bluish-grey back and wings.

Peregrine Falcon
Falco peregrinus

Adult male, February.

Taxonomy Polytypic (19).
Description L 38–51cm
(15–20in). Large thickset falcon
with blackish upperparts;
blackish head, nape and
pronounced malar stripe form
a dark helmet contrasting
with white sides to neck,
throat and breast, abdomen
heavily barred, bill dark with
yellow cere, thick muscular
legs feathered and barred,
long powerful yellow talons.
Migrants show great colour
variation with upperparts
dark blue to dark grey or
black. Juvenile has brownish
upperparts and heavily streaked
breast and underparts. In flight
long tapering pointed wings
usually flexed and appear broad
with barred primaries; narrow
wedge-shaped tail.
Similar species Merlin is
smaller with heavily streaked
underparts and lacks dark
helmet.
Voice Silent.

Habitat and behaviour
Solitary. Forages in all habitats.
Perches to select prey including
on high-rise buildings on
Grand Cayman. Almost
exclusively hunts birds, both
large (tropicbirds, herons,
egrets and ducks) and small
(shorebirds, doves, swallows),
striking prey on take-off, in
the air from stoops (swoops)
from heights up to 1,000m,
or on the ground, showing
extreme manoeuvrability and
acceleration. Patrols the eastern
bluffs of Cayman Brac attacking
Brown Booby adults and

young, characteristically ripping
the breast from the adults.
Range Cosmopolitan. Breeds
from North America to South
America; northern North
American birds migrate to South
America and the West Indies.
Status Fairly common winter
visitor and passage migrant,
September–May. Numbers
increasing and negatively
impacting Brown Booby
population (many reports from
Caribbean islands of impacts
on fragile seabird colonies
has come as a mixed blessing.
Conservation Cites 1.

Juvenile has streaked underparts, barred in adult, February.

Sora
Porzana carolina

Adult has black crown stripe and white streaks on brown back, January.

Taxonomy Monotypic.
Description L 22cm (8.5in).
Small rail. Non-breeding adult has dark brown and black upperparts with white streaks; grey face, neck and breast, short stout yellow bill with greenish tip; abdomen and flanks barred blackish and white, undertail-coverts whitish, pointed tail, olive legs. Breeding adult has black mask and throat to centre breast, this reduced in female and absent in juvenile. Juvenile face and breast are buffy.
Similar species None.
Voice Call whistled *koowe* and short *kieu*.
Habitat and behaviour All inland wetlands with fresh and brackish water preferred. Usually secretive, remaining close to cover; occasionally forages in the open taking plant food, crustaceans, insects and worms.

Range Breeds in North America; winters in the United States on both west and Atlantic/Gulf coasts to central Peru and northern South America, and the West Indies.

Status Fairly common winter visitor on Grand Cayman, and uncommon on Little Cayman and Cayman Brac, September–May; some passage migrants occur in late May.

Adult in breeding plumage shows black crown stripe, face mask and throat, January.

Purple Gallinule
Porphyrula martinica

Taxonomy Monotypic.
Description L 33cm (13in).
Undertail coverts entirely white
in all plumages. Adult has
brilliant iridescent purple head,
neck, breast and underparts,
powder blue frontal shield, red
bill with yellow tip, iridescent
bronze-green mantle and
coverts, bright yellow legs
and feet with long toes, and
pointed tail. Juvenile has
brownish upperparts washed
green, dull brownish bill
and frontal shield, and buffy
underparts.
Similar species Juvenile
Common Gallinule is grey with
white flank stripe. Juvenile
American Coot is greyish with
whitish chin and throat, no
frontal shield.
Voice Cackles and squawks
like a gallinule but notes do
not descend *ka-ka-ka-ke,* and
warning *keip*.
Habitat and behaviour
Clambers in mangrove roots
and walks on dense pond
vegetation and pond edges.
Mainly takes plant material;
also crustaceans, insects, larvae

Adult, September.

and worms. Cocks and flicks tail
while walking. Breeding May–
October; resident population
is small and secretive, staying
close to cover, preferring
herbaceous wetlands. Bulky
cup nest of plant vegetation
above water, often among
Buttonwood roots or Cattails,

3–12 buffy eggs spotted, 2–5
chicks usually fed by both
parents. Migrants, often in
loose groups, very visible on
lagoons and ponds, flooded
grasslands, the Ironshore and
urban areas.
Range Breeds from the United
States Atlantic/Gulf coast to
Argentina, and the West Indies
where common and breeding
in Cuba and Hispaniola,
uncommon elsewhere. North
American birds winter south
through the breeding range.
Status Uncommon breeding
resident on Grand Cayman,
population has declined due
to loss of habitat; intermittent
resident and rare breeder on
Little Cayman and Cayman Brac
(no recent records). Uncommon
winter visitor and locally fairly
common to occasionally
abundant passage migrant,
August– mid-July.

Juvenile has greenish wash on brown upperparts, August. Breeds on Grand
Cayman only.

Common Gallinule
Gallinula galeata

Local name Red Seal Coot.
Taxonomy Polytypic (12).
Description L 34cm (13in).
Charcoal-grey rail with brownish back, white flank stripe, two white undertail-covert flashes under black tail in all plumages. Breeding adult has bright red bill tipped with yellow, red frontal shield, and long greenish yellow legs with red tibial bands. Non-breeding adult is browner with dull reddish bill (yellowish tip) and frontal shield; juvenile greyish-brown with paler throat and underparts, dull brown bill (yellowish tip) and frontal shield.
Similar species Juvenile American Coot and Purple Gallinule lack white line along flanks.
Voice Single *kik,* cackles, and loud *ke-ke-kekrra-kru-kru-kru* descending.

Adult breeding showing bright red frontal shield, bill and tibial bands, April.

Habitat and behaviour All wetland habitats with fresh to brackish water preferred, also shores, and gardens with ponds; rafts of migrants winter on saline lagoons. Mainly herbivorous taking terrestrial and aquatic plant material; also insects, molluscs, larvae, spiders and worms. Unafraid of humans, coming to share food with domestic ducks. Shaped like young domestic chicken on land; flicks tail, and moves head back and forth when swimming. Bulky platform nest of sticks, leaves and roots close to or over water, 3–9 whitish-buff eggs with reddish spots, 1–4 fledglings quickly follow parents to be fed on the water.
Range *G. g. cerceris* is a West Indian endemic subspecies. North American birds, probably *G. g. cachinnans* winter throughout the Greater and Lesser Antilles.
Status Abundant resident on Grand Cayman, breeding throughout the year. Fairly common to uncommon breeder on Little Cayman and fairly common in limited habitat on Cayman Brac. An abundant winter visitor, August–March.

Adult non-breeding has reddish frontal shield and base of bill, November.

Juvenile has whitish flank stripe, and dull shield and proximal part of bill, July.

American Coot
Fulica americana

Taxonomy Polytypic (2).
Description L 38–40cm (15–16in). Blackish head and neck, dark slate-grey body, small dark red dot on centre of frontal shield, thick white bill with faint incomplete ring near tip, greenish legs with large lobed toes, outer undertail-coverts white. White-tipped secondaries show as white trailing edge to wing in flight. Juvenile is greyish with pale throat and brownish bill. Fledglings blackish with red on head neck and throat.
Similar species Juvenile Common Gallinule has white stripe on flanks. Juvenile Purple Gallinule has brownish upperparts washed green, dull brownish bill and frontal shield, and buffy underparts.
Voice *Took* repeated; clucks and grating cackle.
Habitat and behaviour All wetland habitats with fresh to brackish water preferred when breeding, including flooded

Adult has partial ring around bill tip and large feet with lobed toes, December.

marl pits on Grand Cayman. Rafts of migrants winter on saline lagoons. Breeding season extends throughout the year

Adult has red dot in centre of frontal shield and white flashes on outer tail-coverts, January.

Juvenile has greyish upperparts, whitish underparts and brownish bill.

on Grand Cayman, mainly March–August on Cayman Brac and Little Cayman. Bulky nest platform floating on or over water often concealed in Cattails, clutch 2–12 brownish eggs, spotted brown and black, 1–3 young soon follow parents to be fed on the water. Dives for food, mainly plants; also takes aquatic and terrestrial larvae, worms and other invertebrates, and grain fed to domestic fowl.
Range *F. a. americana* breeds from North America to Costa Rica and the West Indies in the Bahamas, Greater Antilles and Cayman Islands; winters south throughout the breeding range.
Status Fairly common breeding resident on Grand Cayman and Cayman Brac (although absent in some years), uncommon and irregular on Little Cayman; abundant winter visitor, August–late May.

Limpkin
Aramus guarauna

Taxonomy Polytypic (4).
Description L 69cm (27in).
Large solitary wading bird.
Adult has dark brown plumage marked with white streaks and triangles from head to mid-body, long neck; long heavy, slightly decurved bill and long olive legs. Immature paler and more spotted.
Similar species Immature night herons have shorter straight bills and shorter legs. Immature Glossy Ibis has thinner, more decurved bill and dark unstreaked plumage.
Similar species None.
Habitat and behaviour Urban gardens, herbaceous wetlands; walks deliberately flicking tail.
Range *A. g. pictus* breeds in Florida and the West Indies in the Bahamas, Cuba and Jamaica, *A. g. elucus* breeds in Hispaniola and Puerto Rico, a third subspecies breeds in Middle America and a fourth in South America.
Status Vagrant to Grand Cayman, April–May.

Adult has white streaks on head and neck forming triangles on body and wings.

Black-bellied Plover
Pluvialis squatarola

Taxonomy Monotypic. Also known as Grey Plover.
Description L 26–34cm (10–13.5in). Non-breeding adult and immature have brownish-grey upperparts, prominent black eye, dark ear-coverts, whitish supercilium, whitish underparts lightly streaked on breast, short heavy black bill, and dark legs. Juvenile has more pronounced streaking on underparts. Birds

Adult in breeding plumage, May.

Adult. Transition to winter plumage, October.

in early fall and late spring show black mottling on face and underparts. Breeding (April–August) adult has pale grey crown, white band from forehead and supercilium down sides of neck to shoulder, black face and breast to mid-abdomen, upperparts mottled black and pale grey. In flight shows black axillaries on underwing, broad white band across upperwing, white rump and undertail-coverts, whitish tail finely barred distally.

Similar species Non-breeding American Golden Plover is smaller and browner, in flight upper tail is dark and black axillaries are absent; breeding has black undertail-coverts.

Voice Plaintive *klee-u-ee,* and single *klee.*

Habitat and behaviour Mainly coastal on fringing reefs, shores, saline lagoons, also on inland ponds, flooded marl pit edges, wet grassland, and herbaceous wetlands.

Range Holarctic, wintering to southern hemisphere. Breeds in North American Arctic; winters on both coasts from North America to Chile and Argentina, and the West Indies.

Status Fairly common winter visitor and locally common passage migrant, July–May; a few immature over-summer.

Adult non-breeding has brownish-grey crown, and upperparts spotted whitish on feather edges, February.

Juvenile. Note well-defined white edges to feathers on upperparts, streaked breast and flanks, October.

Wilson's Plover
Charadrius wilsonia

Taxonomy Polytypic (3).
Description L 18–20cm
(7–8in). Small plover with large
head. Non-breeding adult
and breeding female have
large thick black bill, greyish-
brown crown and upperparts,
white of collar and forehead
continuous with supercilium to
behind eye, white underparts,
wide brown breast-band may
appear incomplete, greyish-
pink legs. Breeding male has
black forecrown bar, lores
and complete breast-band.
Breast-band brown in female.
Juvenile has breast-band
mottled with buff, upperparts
edged pale, and dull yellowish
legs. In flight shows dark outer
upperwing with broad white
band.
Similar species Semipalmated
Plover has narrow breast-band,
very short bill and yellow-
orange legs.
Voice Usually silent, sharp

Male in breeding plumage has black lore, forecrown and breast-band, February.

wheet call.
Habitat and behaviour The
Ironshore, sandy shores and
saline lagoon edges, prefers
drier areas than Semipalmated
Plover, usually singly or in
very small groups with other
waders, foraging for crabs,
insects, molluscs and worms.
Range *C. w. wilsonia* breeds
from North America's Pacific
and Atlantic-Gulf coast to
Mexico and Belize and the
West Indies in the Bahamas,
Greater Antilles and Lesser
Antilles to Dominica and
Grenada. Winters throughout
the breeding range.
Status Uncommon to locally
fairly common, short-stay
winter visitor and passage
migrant in all months, rare
from mid-June to mid-July. The
Cayman Islands are among
the few islands in the northern
West Indies without breeding
populations.

Adult non-breeding and juveniles are similar with brown breast-band, February.

Semipalmated Plover
Charadrius semipalmatus

Adult male in breeding plumage, May.

Taxonomy Monotypic.
Description L 18.5cm (7.25in).
Non-breeding adult has dark
brown crown and upperparts;
white forehead, supercilium and
collar, dark ear coverts, narrow
dark brown breast-band, very
short blackish bill with small
amount of orange at base,
and yellow legs. Breeding male
has black frontal bar and black
patch from eye to ear coverts,
narrow black breast-band forms
a black collar, bill orange with
black tip, legs orange-yellow;
breast-band and collar dark

brown in female. Juvenile similar
to non-breeding; upperparts are
edged pale, breast-band may
be reduced, bill all-black, and
legs dull yellow. In flight shows
broad white band along dark
outer wing.
Similar species Wilson's Plover
has wider breast-band, heavier
longer black bill and grey-pink
legs.
Voice High pitched *koo-wee* or
kweet.
Habitat and behaviour
Beaches, lagoon edges,
disturbed habitats near water,

less frequently on herbaceous
wetlands. Forages for worms,
grasshoppers, beetles and
ants; alternates quick runs with
pauses.
Range Breeds in northern North
America; winters from United
States Pacific and Atlantic-
Gulf coasts to southern South
America, and the West Indies.
Non-breeding birds often
summer in the wintering range
to Panama.
Status Fairly common short-stay
winter visitor and passage
migrant, July–June.

Adult non-breeding, October.

Juvenile has black bill, November.

Killdeer
Charadrius vociferus

Adult. The only plover to have two breast-bands, May.

Taxonomy Polytypic (3).
Description L 25cm (10in).
Large, long-bodied plover
with dark brown head and
back, orange eye-ring, white
on forehead continuing below
eye, white supercilium and
collar, two black breast-bands
on white underparts (anterior
encircles the neck), long black
bill and pinkish legs. Juvenile
is similar but duller with
upperparts edged pale. In
flight shows orange rump and
uppertail-coverts, blackish tail
edged white, and broad white
wing-stripe across black outer
upperwing.
Similar species None. Only
plover with two breast bands
and orange rump.
Voice Call *kil-deer* and *kdee-
dee-dee,* alarm *deet* repeated.
Habitat and behaviour Forages
for grasshoppers, flies and
beetles, seeds on herbaceous
wetlands, wet grassland and,

rarely, saline lagoon edges.
Often solitary, feeding and
calling all night.
Range *C. v. vociferus* breeds
from North America to Mexico
and winters to South America
and the West Indies. *C. v.
ternominatus* breeds in the

Greater Antilles and Bahamas
and is sedentary. A third
subspecies occurs in South
America.
Status Uncommon short-stay
winter visitor and uncommon
to fairly common passage
migrant, September–May.

Adult shows orange rump and uppertail and broad white wing-stripe.

Black-necked Stilt
Himantopus mexicanus

Local name Tell-tale, Soldier.
Taxonomy Polytypic (2).
Description L 34–39cm (13.5–15.5in). Large, slim wader with long neck, long flexible, needle-like bill and extremely long salmon-pink legs. Adult is white with black crown and hindneck, white patch over eye; back and wings are black in adult male, brownish-black in female. Juvenile has greyish-brown upperparts edged pale, bill pinkish at base, and yellowish legs. In flight, black and white with greyish tail and long trailing pink legs.
Similar species None.
Voice Noisy when disturbed, hence local names, alarm call continuous *yip-yip-yip* while circling, also *kr-eik* when foraging at night.
Habitat and behaviour Solitary pairs and small groups occur on all inland

Adult male has a black back, April.

wetland habitats; fresh and brackish preferred, including golf courses, spoil banks and garden ponds; larger colonies breed on saline lagoon edges, March–July. Nest is a scrape

Adult female has brownish-black back, December.

with varying amounts of vegetation close to water's edge or on Ironshore outcrops in lagoons, 2–5 buffy eggs with dark markings and spots. Downy long-legged fledglings are fed and defended by highly territorial parents who perform 'broken-wing' distraction behaviour. Preyed upon by raptors, barn owl, cats and dogs. Takes mainly aquatic insects, crustaceans, worms and water plant seeds.

Range Americas. *H. m. mexicanus* breeds from North America on the Atlantic-Gulf coast and south-west interior to southern South America, and the West Indies from the Bahamas south to Antigua. Winters south through the breeding range.

Status Abundant breeding resident, Grand Cayman; rare to fairly common in winter, common March–August, Little Cayman and Cayman Brac, when flocks arrive to

Courtship. Female with brownish-black back (right), March.

breed, after which the majority migrate. It is not known if those joining the small sedentary resident population on the latter are from Grand Cayman or North America.

Juvenile. Note yellowish legs, pinkish base to bill and indistinct patch over eye, September.

Downy young, April.

American Avocet
Recurvirostra americana

Adult, non-breeding, has pale grey head and neck, October.

Taxonomy Monotypic.
Description L 40–51cm (16–20in). Large, slender wader. Non-breeding adult has pearl-grey head and neck, white body, black back and wing-coverts with large white patch, very long greyish-blue legs and long, black, slender upcurved bill; less pronounced in male. Breeding adult has cinnamon head and neck. Juvenile has brown wash on head.
Similar species None.
Voice Silent on migration.
Habitat and behaviour Herbaceous wetlands, including ponds on golf courses and saline lagoons; feeds on aquatic invertebrates.
Range Breeds in western and central North America, and central Mexico; winters to Florida, Mexico and Guatemala, and the Bahamas and Cuba where it is rare; rare on passage or vagrant elsewhere in the Greater Antilles and Cayman Islands.
Status Rare passage migrant, August–May.

Adult transitioning to breeding plumage, April.

Greater Yellowlegs
Tringa melanoleuca

Taxonomy Monotypic.
Description L 33–38cm (13–15in). Large shorebird with white eye-ring, long robust bill (1.5 x head length) slightly upturned and with feathers on lower mandible not reaching nostril, and long yellow legs. Non-breeding adult has brownish-grey upperparts spotted white, streaked head and breast, bill paler at base, white abdomen and undertail-coverts, flanks with faint bars. Breeding adult has black and grey upperparts speckled with white, bill dark; head, neck and breast heavily streaked, flanks barred black. Juvenile upperparts finely spotted with buff. In flight shows dark unmarked upperwing, white rump and upper tail, barred outer tail, and feet extending beyond tail.
Similar species Lesser

Adult non-breeding. Note bill paler at base and feathers on lower mandible do not reach nostril, March.

Yellowlegs is smaller, straight all-dark bill is same length as head and feathers on lower mandible reach nostril, call usually 2 syllables.
Voice 3–4 syllable *tcheu-tcheu-teu* descending; alarm call continuous *teu-teu*.
Habitat and behaviour All wetland habitats, including the Ironshore, fringing reefs and wet grassland, with other waders; more active when foraging than the Lesser Yellowlegs.
Range Breeds in North America; winters from the southern United States to the southern tip of South America, and the West Indies. Birds often over-summer in the northern wintering range.
Status Uncommon to fairly common winter visitor and fairly common to common passage migrant (mainly spring), July–June; a few birds over-summer. Less numerous than Lesser Yellowlegs.

Adult non-breeding, February.

Lesser Yellowlegs
Tringa flavipes

Taxonomy Monotypic.
Description L 25–28cm (9.75–11in). Slim, medium-sized shorebird with thin straight all-dark bill (just longer than head) and yellow legs. Very similar to Greater Yellowlegs but smaller and more slender. Non-breeding adult has upperparts brownish-grey with white spots, head may have darker crown, neck and breast finely streaked. Breeding adult has upperparts blackish heavily speckled with white and buff; head, neck and breast heavily streaked, barring on flanks reduced or absent. Juvenile upperparts are brownish with large whitish spots, bill may have yellow base. In flight shows dark unmarked upperwings, white rump and uppertail-coverts, and feet extending beyond tail.
Similar species Greater

Adult non-breeding. Bill length similar to that of head, April.

First-winter has straight bill, yellow at base and brownish-grey coverts, April.

Yellowlegs is larger, bill is longer, pale at base and usually slightly upturned; call 3–4 syllable and louder.
Voice 1–2 syllable *tu* or *tu-tu* quieter than Greater Yellowlegs, and alarm *kleet* and continuous *teu-teu*.
Habitat and behaviour All wetland habitats, and the Ironshore, fringing reefs, and golf courses.
Range Breeds in northern North America; winters from the southern United States to southern tip of South America, and the West Indies.
Status Fairly common winter visitor and common to locally abundant passage migrant, July–May. It does not over-summer.

Solitary Sandpiper
Tringa solitaria

Taxonomy Polytypic (2).
Description L 19–23cm (7.5–9in). Medium-sized shorebird with medium length fine blackish bill greenish at base, pronounced white eye-ring continuous with white stripe to eye forming 'spectacles', black lores, dark greenish legs. Non-breeding adult has grey-brown upperparts finely spotted white; streaked head, neck and breast, rest of underparts white. Breeding adult is more heavily streaked. Juvenile had brownish wash over head and upperparts. Bounding flight shows dark unmarked upper- and underwing and rump, dark central band on tail with bars and white edges.
Similar species Non-breeding Spotted Sandpiper is smaller, lacks eye-ring, unspotted brown back, legs yellowish and short. Stilt Sandpiper has longer slightly decurved bill and lacks eye-ring.
Voice Call *weet-weet* and

Adult in breeding plumage, April.

weet-weet-weet.
Habitat and behaviour Solitary, except small flocks in migration; walks slowly foraging in herbaceous wetlands, flooded grassland, rainwater pools on roads, also saline lagoons. Raises wings up when alighting and folds them slowly; bobs head

and front of body continually.
Range *T. s. solitaria* breeds in northern North America; winters from Mexico to central South America and the West Indies.
Status Rare winter visitor and fairly common passage migrant, August–May.

Adult non-breeding, March.

Willet
Tringa semipalmata

Willet, eastern subspecies, in breeding plumage, May.

Taxonomy Polytypic (2).
Description L 38–40cm
(15–16in). Large heavy-bodied
shorebird with pale eye-
ring and area above lores,
bluish-grey legs. Non-breeding
adult of eastern and western
subspecies similar: upperparts,
throat and breast pale grey,
rest of underparts whitish.
Breeding adult of eastern
subspecies has long, thick
dark bill paler at base, dark
brown upperparts, heavily
barred underparts and buffy
flanks. In flight shows broad
wings; upperwing has black
flight feathers and broad white
wing-stripe, black underwing-
coverts, white rump and grey
outer tail. Juvenile has grey-
brown upperparts with black
centres, dark crown.
Similar species None have
black and white wing pattern.

Voice Around nest *pill-will-
willet;* alarm call continuous
yip-yip-yip.

Habitat and behaviour All
wetland habitats including
shores. Breeds on saline lagoon

Adult non-breeding, March.

Adult's diagnostic wing pattern shows black flight feather crossed by white wing-stripe, and white rump, June.

edges, May–July, 4 buff eggs heavily mottled; adults are very territorial and show distraction behaviour at nest or near fledglings.

Range Eastern race *T. s. semipalmata* breeds in eastern North America on the Atlantic/ Gulf coast, and in the West Indies in the Bahamas, Greater Antilles, Cayman Islands and Antigua; winters in breeding range to eastern South America. Western race *T. s. inornata* breeds on the western North America, winters to western South America and only occurs in passage in the West Indies.

Status *T. s. semipalmata* is an uncommon breeding resident; both subspecies are fairly common in winter and on passage, mainly in spring.

Juvenile. Note pale stripe above lores, and pale edges to dark-centred wing-coverts, September.

Spotted Sandpiper
Actitis macularius

Adult non-breeding has unspotted underparts and white notch before wing, February.

Taxonomy Monotypic.
Description L 18–20cm (7–8in). Small, short-necked shorebird with black eye-line, long white supercilium; underparts white with white notch before wing. Non-breeding adult has unmarked brown upperparts, brown lateral patches on breast, bill dark olive paler at base, legs yellow. Breeding adult, observed in late spring and early fall, has white underparts with large black spots, pinkish-orange bill with black tip, upperparts olive-brown with scattered black bars, legs pinkish. In flight shows brown rump and tail and white stripe down mid-wing on short wings. Juvenile lacks spots and back and wing-coverts are barred.
Similar species Solitary Sandpiper has longer neck with more extended posture, dark bill, longer greenish legs and lacks white notch on shoulder.
Voice Call *peet* in flight and slow *weet-weet-weet*.

Habitat and behaviour All wetland habitats including the Ironshore. Small loose flocks usual in migration otherwise solitary. Tips hind parts up and down when walking; takes short flights with shallow wing-strokes interspersed with glides and wing-beat is below horizontal.

Range Breeds in North America excluding the Gulf coast; winters from southern United States to central South America and the West Indies, where it also occurs on passage.
Status Fairly common winter visitor and common passage migrant, most numerous in late spring, July–June.

Adult in breeding plumage, April.

Upland Sandpiper
Bartramia longicauda

Taxonomy Monotypic.
Description L 28–32cm (11–13in). Large
shorebird with small head, long thin neck, short
wings, long tail, large dark eye with white eye-ring
in pale buffy face, short yellow bill with black
culmen and tip, and yellow legs. Upperparts
barred brown with pale edging to feathers; head
and neck heavily streaked, breast and flanks buff
with dark chevrons. In flight shows unmarked
wings with pale inner wing and black rump.
Similar species None.
Voice Silent.
Habitat and behaviour Solitary, on wet grassland;
also regularly perches on fences; runs and stops
when foraging like Semipalmated Plover.
Range Breeds in North America; winters in South
America. Passage migrant in the West Indies
where it is very uncommon in the Bahamas,
Puerto Rico and locally in the Lesser Antilles.
Status Rare passage migrant, September–
October and March–May.

Adult has small head, long neck, white eye-ring and
chevrons on breast and flanks.

Whimbrel
Numenius phaeopus

Taxonomy Polytypic (4).
Description L 38–46cm (15–18in). Large
shorebird with long downcurved bill (8–10cm)
paler at base. Dark brown upperparts spotted
with black and buff, dark crown with pale
median stripe, wide pale supercilium, dark
eye-line, finely streaked neck and breast, whitish
underparts with chevrons on flanks, blue-grey
legs and feet. In flight appears grey-brown
overall with barred underwings on pointed
wings. Breeding and juvenile plumages similar.
Similar species Only large brown wader with
downcurved bill.
Voice Rapid whistle *qui-qui-qui.*
Habitat and behaviour Sandy shores, saline
lagoons, and disturbed wetlands, where it
probes for crabs on beaches, and picks insects,
crustaceans, polychaete worms and molluscs.
Usually solitary but small groups occur in migration.
Range Holarctic breeder. *N. p. hudsonicus* breeds
in northern North America; winters coastally
from the United States to South America and the
West Indies where it is uncommon to rare.

Status Uncommon winter visitor and fairly
common passage migrant, August–May;
individuals occasionally over-summer remaining
at the same site for several years.

Adult. Note central crown stripe and chevrons on flanks, May.

Ruddy Turnstone
Arenaria interpres

Taxonomy Polytypic (2).
Description L 21–23cm (8–9in). Stocky, short-legged shorebird with short, black, slightly upturned, pointed bill. Non-breeding adult has dark brown head with white patches, dark breast-band, brownish upperparts edged with buff, white underparts and pale orange legs. Breeding male has back and wing-coverts rusty-orange and black, head and neck white patterned with black, black throat and breast-band, orange-red legs; female duller with brownish breast-band. Juvenile similar to non-breeding adult with upperparts

Adult male, September.

edged widely with buff, partial breast-band and dull yellow

Adult non-breeding, November.

First-winter has pale edges to worn feathers on back and wing-coverts, November.

legs. In flight adult shows rufous, black and white pattern on upperparts and tail, white wing-stripe on upperwing and white underwing.
Similar species None.
Voice Low *keip,* and rattling flight call.
Habitat and behaviour Shores, especially the Ironshore, saline lagoon edges, harbours and inland wetlands; often overwinters at the same site. Turns over stones and vegetation probing for food. Very wide diet includes molluscs and worms; also scavenges in coastal urban areas.
Range Holarctic breeder. *A. i. morinella* breeds in Alaska and Arctic Canada and winters from the United States Pacific and Atlantic-Gulf coasts to South America and the West Indies. Immature birds often remain in the wintering range. *A. i. interpres* breeds in Arctic Canada and Palearctic.
Status in the Cayman Island Fairly common winter visitor and common passage migrant; non-breeders over-summer on Grand Cayman.

Red Knot
Calidris canutus

Adult transitioning to breeding plumage, May.

Taxonomy Polytypic (5).
Description L 25–28cm
(10–11in). Medium-sized,
bulky, round-bodied shorebird
with thick blunt straight bill
and short greenish legs. Non-
breeding adult has greyish
upperparts with scapulars
narrowly edged white, whitish
supercilium; neck and breast
greyish-brown with streaks,
barring on flanks. Juvenile
similar but upperparts scaly.
Breeding adult has crown and
back grey; scapulars patterned
black, cinnamon and buff;
face, neck and underparts
cinnamon-rufous, dark legs.
In flight shows diagnostic grey
rump and tail.
Similar species Winter 'peeps'
and White-rumped Sandpiper
are smaller. Dunlin has longer
bill drooping at tip and black
legs. Short-billed Dowitcher has
very long bill and white rump.
Voice Silent.

Habitat and behaviour Saline
lagoon edges, shores.
Range Holarctic. *C. c. rufa*
breeds in the Canadian Arctic
and *C. c. roselaari* breeds in
Alaska; both winter in South

America. Rare and casual on
passage in the West Indies in
all months.
Status Rare passage migrant,
August–December, and
February–May.

Adult non-breeding has streaks on breast and barring on flanks, March.

Sanderling
Calidris alba

Adult transitioning to breeding plumage, May.

Taxonomy Monotypic.
Description L 18–22cm
(7–8.5in). Small shorebird
with the palest non-breeding
plumage. Non-breeding adult
has silvery grey upperparts,
white face and underparts,
black 'shoulder' (lesser coverts),
white notch before wing, short
black bill and legs; lack of hind
toe is diagnostic. Breeding
adult rufous on head, neck
and breast; mantle and covert
feathers black and rufous with
white edges. In flight shows
wide white wing stripe, black
leading edge to wing, and
white sides to tail. Juvenile has
upperparts speckled black and
white, dark crown and lores,
white underparts, breast may
be buffy.
Similar species Non-breeding
Red Knot is larger and darker,
legs greenish; in flight grey rump
and tail are diagnostic. Spotted
Sandpiper has plain brown
upperparts and greenish legs.
Voice Call *wick;* high *ti-ti-ti*
among flocks feeding.
Habitat and behaviour Flocks
make rapid synchronised runs
along sea edge beachcombing
for worms, small molluscs and
crustaceans; also on lagoon
edges and golf courses.

Range Holarctic breeder. Breeds
in Canadian Arctic; winters on
both coasts of the Americas
and West Indies; some summer
in the wintering range.
Status Uncommon to locally
common short-stay passage
migrant and winter visitor,
July–June.

Adult non-breeding. The palest small sandpiper observed in winter, December.

Semipalmated Sandpiper
Calidris pusilla

Adult non-breeding plumage, February.

Taxonomy Monotypic.
Description L 14–16.5cm (5.5–6.5in). Very small shorebird, one of three difficult to distinguish species of 'peeps'. Non-breeding adult has greyish-brown upperparts, dark crown and ear coverts, whitish supercilium, white underparts with lateral streaking on breast, straight black bill broad at base (length overlaps with Western Sandpiper), partly webbed toes and black legs. Breeding adult has greyish-brown upperparts with black centres and pale edges to scapulars (may show rufous), white underparts with darkly streaked neck and sides of breast, streaks usually not extending to flanks. Juvenile variable; has white supercilium, dark ear coverts, bright greyish-buff or rufous tinged crown, ear coverts and scapulars; back and wings have black centres edged pale giving a scaly effect.
Similar species Least Sandpiper has finer, slightly decurved bill and greenish-yellow legs.

Western Sandpiper is paler but similar; bill usually thicker at base, longer and slightly decurved, legs longer, juvenile and breeding plumages show bright rufous on scapulars.
Voice Call loud short *kre-ip*.
Habitat and behaviour Sandy shores, saline lagoons, seldom on herbaceous wetlands or rain pools; probes for larvae, worms and molluscs.
Range Breeds in northern North America; winters from Florida and Yucatan Peninsula to both coasts of South America, and the West Indies: where it occurs in all months.
Status Uncommon to locally common short-stay winter visitor and passage migrant in all months; numbers have declined in the 21st century. Occasionally absent for entire months from one or more of the Islands.

Adult non-breeding. Note dark cap, long supercilium and straight bill, May.

Western Sandpiper
Calidris mauri

Adult non-breeding. Bill is wider at base and droops slightly at tapering tip, March.

Taxonomy Monotypic.
Description L 15–18cm
(5.75–7in). Almost identical
to Semipalmated but slightly
larger with bill drooped at tip
– most noticeably in females –
with thinner tip and thicker at
base. Non-breeding adult has
pale grey upperparts, white
underparts with pale greyish
breast sides (seldom a full
breast-band), black bill and legs.
Breeding adult has bright rufous
on crown, upperparts and ear
coverts, white V on scapulars,
white underparts heavily spotted
on breast and chevrons along
flanks. Juvenile is greyish with
rufous scapulars, dark lores,
white underparts with faint
streaking on sides of breast.
Similar species Semipalmated
Sandpiper difficult to
separate: has straight bill,
usually shorter, juvenile and
breeding plumages lack rufous
on scapulars. Least Sandpiper
is browner, breast heavily
streaked brown, greenish-
yellow legs.

Voice Usually silent, call *chit*.
Habitat and behaviour
Shores, saline lagoons and
disturbed mangrove wetlands;
often submerges head when
searching for prey in water.
Range Breeds around the
Bering Strait in Alaska (and
in Siberia); winters from both
coasts of the United States to
northern South America, and
the West Indies, where it is
uncommon and occurs in all
months.
Status Uncommon passage
migrant, August–May,
often solitary and probably
overlooked when in flocks with
other 'peeps'. Migrants arrive in
juvenile or winter plumages.

Juvenile. Feathers on scapulars are edged rufous, September.

Least Sandpiper
Calidris minutilla

Adult non-breeding. Yellowish legs shown in all plumages are diagnostic, October.

Taxonomy Monotypic.

Description L 12.5–16.5cm (5–6.5in). Smallest 'peep'. Non-breeding adult has greyish-brown upperparts and white underparts, entire breast washed or streaked brownish-grey, bill slightly drooped at finely pointed tip; greenish-yellow legs are diagnostic. Breeding adult has dark brown mantle and scapulars, with black centres edged rufous or white, dark crown and ear coverts, darkly streaked head and breast. Juvenile (fresh) is lighter brown with rufous on upperparts, ear coverts and crown; white supercilia meet above bill.

Similar species No other small sandpiper has greenish-yellow legs. Non-breeding Semipalmated and Western Sandpipers have blackish legs.

Voice Flight call high *treeep*.

Habitat and behaviour All wetland habitats with fresh and brackish water preferred, especially temporary rain pools and small inland ponds; remains at water's edge, seldom wading; hunched appearance.

Range Breeds in northern North America; winters from the United States to central South America and the West Indies. Non-breeding birds summer in the wintering range.

Status Fairly common winter visitor and common passage migrant, July–June. Juveniles common in July and August.

Adult transitioning to breeding plumage with white tips to scapulars, and back and wings are edged rufous and white, May.

White-rumped Sandpiper
Calidris fuscicollis

Majority on spring passage are in transition to breeding plumage, April. Note wings project beyond tail tip.

Taxonomy Monotypic.
Description L 18–20cm (7–8in). Small wader, slightly larger than peeps, has white rump in all plumages. Non-breeding adult has greyish-brown upperparts edged whitish, white supercilium, streaked greyish head, breast and flanks, wingtips projecting beyond tip of tail and black legs. Medium length straight black bill with small red or orange spot at base of lower mandible is diagnostic but spot hard to see. Breeding adult has upperparts heavily spotted with grey and black, rufous on crown, ear coverts and scapulars; heavily streaked breast and flanks. Juvenile has bright rufous and buff on upperparts, rufous on crown and ear-coverts, long white supercilium. In flight shows white rump and uppertail-coverts (only small sandpiper with white rump) and dark tail.
Similar species Semipalmated, Least and Western Sandpipers are smaller with dark rumps and wings that do not extend beyond tail. Baird's Sandpiper (vagrant) has a dark rump.
Voice Call thin insect-like *jeet*.
Habitat and behaviour Wades in deep water in lagoons and ponds, flooded marl pits on Grand Cayman, disturbed wetlands, wet grassland including golf courses.
Range Breeds in the North American Arctic; winters in southern South America to Cape Horn. Uncommon on passage in the West Indies in fall, and very uncommon in spring, except in the Cayman Islands and western Cuba, when the main flyway is through Middle America.
Status Locally common passage migrant in late spring and uncommon in fall, September–October and April–June.

Adult. Orange spot at base of bill, and bill drooped at tip are diagnostic field marks.

Pectoral Sandpiper
Calidris melanotos

Taxonomy Monotypic.
Description L 20–24cm (8–9.5in). Heavy-bodied medium-sized shorebird with heavy streaks on breast that end abruptly at white abdomen in all plumages. Non-breeding adult has upperparts brownish with dark centres and buff edges, white supercilium, black lores, tapered bill yellowish at base droops at tip, short greenish-yellow legs. Breeding adult has upperparts dark brown edged with rufous and buff. Juvenile similar but brighter; rufous on crown, ear coverts, scapulars and tertials.
Similar species Least Sandpiper smaller, bill black, brown breast is not sharply demarcated from white abdomen.
Voice Flight call *jrrrrrt*.
Habitat and behaviour Edges of saline lagoons, Herbaceous wetlands, wet grassland including golf courses, and the Ironshore; seldom forages in water.
Range Breeds in the North American Arctic (and Siberia);

Adult. Heavy streaks ending in centre of breast contrast with white abdomen, May.

winters in southern South America. Passage migrant through the West Indies.
Status Passage migrant, fairly common to locally abundant in fall, August–October, and uncommon in spring, late April–late May.

First-winter has rufous on crown, and edges of scapulars and tertials, January.

Dunlin
Calidris alpina

Taxonomy Polytypic (9).
Description L 20–23cm (8–9in). Stocky medium-sized shorebird with relatively long black bill drooping at tip, and black legs. Non-breeding adult has brownish-grey head and upperparts, brownish breast, white underparts. First-winter similar. Breeding adult has rufous on upperparts and crown, black patch on abdomen, grey streaked head and breast.
Similar Species Red Knot larger with shorter straight bill and greenish-olive legs. Western and Semipalmated Sandpipers are smaller with shorter bills.
Voice Silent.
Habitat and behaviour Rapid probing on beaches, saline lagoon edges and disturbed wetlands.
Range Holarctic breeder. *C. a. hudsonia* breeds in northern

Adult transitioning to breeding plumage shows black on belly, May.

Canada; winters on the United States Atlantic-Gulf-Caribbean coast to the Yucatan Peninsula, and rarely in the West Indies: in the Bahamas and Greater Antilles (absent Hispaniola), Cayman Islands and Lesser Antilles. *C. a. pacifica* breeds in Alaska.
Status Rare migrant or vagrant, September–May.

Adult non-breeding, February.

Stilt Sandpiper
Calidris himantopus

Taxonomy Monotypic.
Description L 20–22cm
(8–8.5in). Slender, medium-
sized shorebird with long bill
thicker at base and drooping at
tip, long wide white supercilium
and long yellowish-green legs.
Non-breeding adult has pale
brownish-grey upperparts
with pale feather edges, white
underparts with faint streaks
on neck and breast, flanks
unbarred. Breeding adult (seen
in spring) has rufous crown and
ear coverts, upperparts dark
brown marked heavily with
rufous and white, neck and
upper breast darkly streaked,
rest of underparts heavily barred
dark brown including flanks and
tail. Juvenile upperparts appear
scaly. In flight shows unmarked
upperwing, white rump and
upper tail-coverts; tail greyish
with dark centre; feet extend
beyond tail.

Adult moulting from breeding to winter plumage, August.

Similar species Non-breeding
dowitchers are larger, very
long straight bills; with white
of upper rump continues as
a wedge up back in flight.
Both yellowlegs are larger with
straight bills and yellow legs.
Dunlin is stocky with black
legs.
Voice *Ke-wee* and ki-wee-wee,
and soft *joof*.
Habitat and behaviour Saline
lagoons, herbaceous wetlands,
flooded grassland, golf course
ponds; forages in deep water,
moving bill up and down or
from side to side.
Range Breeds in the North
American Arctic; winters in
central South America. On
passage in the West Indies,
locally common in the Greater
Antilles in all months.
Status Uncommon to locally
common but irregular on
passage in fall, August–
October, and uncommon in
spring, February–May.

Adult non-breeding has white supercilium and long bill with tip slightly
drooped, March.

Short-billed Dowitcher
Limnodromus griseus

Taxonomy Polytypic (3).
Description L 26–30cm (10–
12in). Medium-sized, stocky
wader with very long, straight,
heavy, blackish bill paler at
base, that droops slightly from
outer third of length; primaries
extend beyond tail at rest. Non-
breeding adult has greyish-
brown barred upperparts
with broad buff edges, white
supercilium, dark-eyeline, fine
streaking on breast does not
extend to abdomen; pale chin,
sides and flanks, undertail-
coverts spotted and barred
brown, legs greenish-yellow.
Breeding adult has upperparts
with black centres to feathers
edged pale and rufous, head
streaked with cinnamon wash,
neck and breast cinnamon-red
with black spots (late spring),
sides and flanks heavily barred,
abdomen white. Juvenile has
crown, upperparts and tertials
with broad rufous-buff edges,
and cinnamon-buff wash
on breast. In flight shows
diagnostic white upper rump as
a wedge between wings, pale
secondaries, and tail barred

Adult non-breeding. Streaking on breast does not extend to abdomen, October.

black and white with black bars
usually narrower.
Similar species Non-breeding
Long-billed Dowitcher is
almost identical, but bill usually
longer (but there is overlap),
grey on breast extensive and
demarcated from abdomen,
wings do not extend beyond
tail. Only reliable separation in
the field is by flight call.

Voice Fluid, short 3-syllable call
tlututu.
Habitat and behaviour Feeds
in small flocks, with up and
down sewing motion of
head while probing mud for
polychaete worms on saline
lagoons and herbaceous and
disturbed wetlands.
Range *L. g. hendersoni* and *L.
g. griseus* breed in northern
North America; winter from
the United States to Brazil and
Peru, and the West Indies,
where *L. g. griseus* is common
in the Bahamas and Greater
Antilles and Cayman Islands; *L.
g. hendersoni* may also occur
in the Cayman Islands due
to its westerly position in the
Caribbean. *L. g. caurinus* breeds
in western North America.
Status Fairly common to
very common winter visitor
and passage migrant, mainly
July–May, but occurs in all
months. Fall migrants arrive
from early July with adults in
faded breeding plumage.

Adult non-breeding. Note white upper rump, October.

Long-billed Dowitcher
Limnodromus scolopaceus

Taxonomy Monotypic.
Description L 28–32cm (11–12.5in). Medium-sized stocky wader, with very long straight heavy blackish bill, paler at base (1.5–2 times head length), white supercilium, dark eye-line. Non-breeding adult has grey-brown upperparts finely edged buff, greyish chin, throat and breast evenly washed with soft grey streaking and demarcated from white abdomen; sides and flanks barred buff, undertail-coverts spotted brown, tail barred black and white, greenish-yellow legs. Breeding adult has upperparts and scapulars with black centres edged with white and rufous, head streaked with cinnamon wash, entire underparts cinnamon-rufous barred with dark brown and white, sides and flanks heavily barred. Juvenile has duller upperparts and tertials edged warm buff. In flight shows white upper rump as a wedge between wings, pale secondaries and dark barred tail.
Similar Species Non-breeding birds are mostly indistinguishable from Short-billed Dowitcher which is smaller, slimmer, with longer

Adult non-breeding has darker centres to back feathers than Short-billed and even grey wash over breast, January.

wings and shorter legs, bill usually shorter (but there is overlap), breast less streaked, flanks and tail more spotted than barred; only reliably separated by flight call.
Voice Flight call light *keek*, singly or in fast series.
Habitat and behaviour Herbaceous wetlands and brackish ponds.

Range Breeds in western American Arctic (and Eurasia); winters from the southern United States to Mexico and, rarely, in the West Indies in the Greater Antilles, and St Kitts and Barbados in the Lesser Antilles.
Status Rare to uncommon, irregular winter visitor and passage migrant, September–May.

Wilson's Snipe
Gallinago delicata

Taxonomy Monotypic.
Description L 27–29cm (10.5–11.5in). Stocky, short-legged wader with exceptionally long straight bill and yellow legs. Adult has dark brown and buff gold stripes on crown and face continuing along dark brown back, buffy breast finely speckled, sides and flanks barred, white abdomen, spotted undertail-coverts. Juvenile similar. In flight shows short orange-tipped

tail and short wings with dark underwings.
Similar species None.
Voice Alarm call harsh *kcaaap* when flushed.
Habitat and behaviour Herbaceous wetlands, flooded

grassland; probes for insects and larvae; in migration on shores and urban/littoral areas. Freezes or explodes from cover into a zigzagging flight on rattling wings.

Range Breeds in North America; winters in United States to northern South America and the West Indies in the Greater Antilles and Cayman Islands.

Status Uncommon winter visitor and locally common passage migrant on Grand Cayman, October–April; uncommon on Cayman Brac and Little Cayman.

Adult Wilson's Snipe, February.

Wilson's Phalarope
Phalaropus tricolor

Taxonomy Monotypic.
Description L 23cm (9in). Slender round body, small head and thin needle-like bill. Non-breeding adult and older juvenile appears whitish in the field with very pale grey upperparts and sides, white face with pale eye-line and white underparts, and yellow legs. Breeding female has grey crown, nape and back, wide black stripe through eye and sides of neck to shoulder, orange fore-neck and breast, white underparts and black legs. Breeding male is paler with brownish-grey upperparts and reddish wash on neck. In flight shows white rump and uppertail-coverts and unmarked upperwing.
Similar species None.
Voice Silent.
Habitat and behaviour Herbaceous wetlands, rain pools in agricultural land, saline lagoons. Agitated active feeding behaviour in water,

Adult in non-breeding plumage.

dashing, crouching, and spinning in circles to stir up prey taken from the surface, also feeds with head and neck submerged, and on land.
Range Breeds in North America; winters in South America. Rare on passage in the West Indies in the Greater Antilles and Cayman Islands.
Status Rare passage migrant mainly on Grand Cayman, September–November, and March–May, one record in winter.

Laughing Gull
Leucophaeus atricilla

Taxonomy Polytypic (2).

Description L 38–43cm (15–17in). Non-breeding adult has hindcrown streaked greyish-brown; white face, underparts and tail, dark grey mantle with black outer primaries spotted white, long black bill drooping at tip, and blackish legs. Breeding adult (May) has black hood to nape, two white crescents above and below eye, dark red bill and dark legs. First-winter greyish-brown on crown, grey mantle and wings with greyish-brown coverts, grey on sides of breast, white

Adult in breeding plumage, May.

Adult non-breeding has white spots on primaries and extent of grey on head varies, February.

First-winter has greyish-brown head and breast and upperwing- coverts brownish, February.

tail with broad dark terminal band. Second-winter similar to non-breeding adult; lacks white tips to primaries.

Similar species Vagrant Bonaparte's Gull is small and tern-like, with small black spot on ear coverts and pinkish legs.

Voice Short *che-waa,* and long loud *ha-ha-ha-ha* series in flocks.

Habitat and behaviour Harbours, marine sounds, saline lagoons.

Range *L. a. atricilla* breeds in the West Indies in large colonies in the Bahamas, Cuba, Virgin Islands, Puerto Rico and Anguilla and smaller colonies elsewhere. *L. a. megalopterus* breeds on United States Pacific and Atlantic-Gulf coasts to Panama and probably occurs in migration. Both subspecies winter in South America and are rare in the West Indies in winter, except in Cuba.

Status *L. a. atricilla* is irregular in winter and locally common in passage, mainly March–November, but occurs in all months.

Ring-billed Gull
Larus delawarensis

Adult breeding. Note white spot on primaries, and bill with black ring and yellow tip, February.

Taxonomy Monotypic.
Description L 46–51cm (18–20in). Non-breeding adult has very fine brown streaks on crown and nape, pale grey mantle with black tips to primaries, slender yellow bill with black subterminal ring, and yellowish legs. In flight pale primaries are black-tipped with small white spots on outer tips, and white tail. Breeding adult has pure white head, bright yellow iris with red orbital ring, red gape, black ring on bill. First-winter has speckling on head, neck, back and breast, grey mantle, brownish upperwing-coverts, bill flesh with black tip, legs flesh; shows dark unspotted outer primaries in flight, dark terminal band on tail. Adult plumage attained in third year.
Similar species Herring Gull larger with pink legs, red spot on lower mandible and ring at tip of bill is never complete.
Voice Silent.
Habitat and behaviour Fringing reefs, marine sounds

First-winter has bill black terminally with pale tip, brownish wing-coverts and no white spots on primaries, February.

and saline lagoons. Usually solitary.
Range Breeds in North America; winters from the United States, Middle America to Panama, and the West Indies in the Bahamas, Cuba and Puerto Rico, rare elsewhere.
Status Rare winter visitor and passage migrant, December–May.

Second-winter has full ring around bill, dark outer primaries and terminal tail-band, December.

Herring Gull
Larus argentatus

Adult non-breeding has speckled head and neck, bill dull with red spot, and pink legs, February.

Taxonomy Polytypic (4).
Description L 56–66cm
(22–26in). Largest gull.
Non-breeding adult has fine
streaking on head and neck,
grey mantle and coverts with
black tips to primaries, heavy
yellow bill with red spot on
lower mandible, pink legs.
Outer two primaries show
white subterminal 'mirror'
spots on black tips in flight,
tail white. Breeding adult has
pure white head and neck and
pale iris. First-winter is entirely
speckled and streaked greyish-
brown, paler on head, bill
blackish or blackish-pink with
black tip, brown iris, pink legs;
in flight brownish overall with
dark brown flight feathers and
tail. Second-winter is greyer
with pale grey inner primaries
on upperwing, greyish-white
rump and black terminal band
to tail. Adult plumage attained
in fourth year.
Similar species Ring-billed

Gull is smaller, bill has black
subterminal ring and yellow
legs.
Voice Silent.
Habitat and behaviour
Inshore waters, harbours.
Range Holarctic. *L. a.
smithsonianus* breeds in North
America; winters mainly in

North America to Panama and
the West Indies, uncommon
in the Bahamas, Cuba and
Cayman Islands, rare elsewhere.
Status Uncommon and
irregular winter visitor and
passage migrant, October–May;
the majority are first- and
second-winter birds.

First-winter plumage mottled brownish-grey, and retains most of black bill of
juvenile, January.

Gull-billed Tern
Gelochelidon nilotica

Adult in breeding plumage, May.

Taxonomy Polytypic (6).
Description L 33–38cm
(13–15in). Heavy-bodied tern
with thick, short, gull-like, black
bill, long black legs, silvery-
white upperparts and shallow
notched tail. Non-breeding
adult has fine speckling on head
with blackish patch behind
eye; in flight appear white
with blackish trailing edge to
long curving outer primaries.
Breeding adult has black crown
to eye and nape. First-winter
has brownish streaks on mantle
and scapulars; in flight shows
dark tips to outer primaries.
Similar species All other
medium-sized black-capped
terns have longer, thinner bills
and tail more forked.
Voice Usually silent, alarm
kee-wak.
Habitat and behaviour Usually
solitary on marine sounds,
flooded marl pits,
wet grassland, also saline
lagoons. Flies like a gull with
slow wingbeats, picks prey
(aquatic and land insects, and
lizards) from surface, does not
dive.
Range Cosmopolitan. *G.
n. aranea* breeds in North
America, Mexican Gulf coast,
and the West Indies in small
colonies (totally 250 pairs) in
the Bahamas, Cuba, Turks and
Caicos Islands; winters from
Middle America and West Indies
to South America.
Status Uncommon migrant,
mainly March–October, but
occurring in all months.

Adult non-breeding has fine streaks on head and black patch behind eye,
January.

Caspian Tern
Hydroprogne caspia

Adult breeding plumage. Black cap almost extends to base of bill, September.

Taxonomy Monotypic.
Description L 48–58cm (19–23in). Largest tern, thickset with heavy red bill tipped blackish, grey mantle, short white tail slightly forked, and black legs. Non-breeding adult has crown streaked black and white, darker behind eye; in flight adult shows diagnostic dark outer third of underwing and dark trailing edge to outer primaries on upperwing. Breeding adult has black crown and darker red bill. First-winter has crown like winter adult and brownish-grey scaling on mantle and coverts.
Similar species Non-breeding Royal Tern has white fore-crown and black hind-crown. In flight underwing shows outer primaries with just a dark trailing edge, darker primaries on upperwing and deeply forked tail.
Voice Silent.
Habitat and behaviour Solitary. Mangrove lagoons, inshore waters and marine sounds. Strong, shallow wingbeats on broad wings; hovers and plunge-dives for fish.
Range Near-cosmopolitan. Breeds in North America and locally on the Atlantic/Gulf coast and western Mexico; winters mainly in southern United States breeding range and Mexico, rarely to southern Middle America and northern Colombia, and the West Indies in the Bahamas and western Greater Antilles (common in Cuba), and very rare to vagrant elsewhere.
Status Rare winter visitor and passage migrant, mainly August–April, but may occur in all months.

Adult non-breeding has dark streaks on crown and dark outer third to underwing.

Royal Tern
Thalasseus maximus

Local Name Sprat bird, Old Tom.
Taxonomy Polytypic (2).
Description L 46–53cm (18–21in). Large tern. Non-breeding adult (observed all year) has white forehead and black hind-crown from behind eye, pearl-grey mantle, white underparts and tail with moderate fork, large bright orange bill, and black legs; in flight shows dark trailing edge to outer primaries below and white tail. Breeding adult (April–July) has black crown to below eye and hind crest and orange-red bill. Juvenile and first-winter have head similar to non-breeding adult with yellow bill and legs, juvenile has brown streaking on mantle and scapulars and dark carpal bar. In flight first-winter upperwing shows dark band on secondaries and dark primaries, tail has black outer edges.
Similar species Caspian Tern has streaked forehead, heavier red bill, less forked tail and dark outer primaries on underwing.
Voice Call high *kri-ii-iip*.
Habitat and behaviour
Inshore waters, marine sounds, fringing reefs and harbours. Fast direct flight on long wings, plunge-dives for fish in inshore shallows, and feeds by

Adult non-breeding has black hindcrown, December.

dipping along the shore. Most frequently seen tern perched on docks or buoys. Begging juveniles may be fed by adults.
Range Americas and West Africa. *T. m. maximus* breeds in North America from Pacific and Atlantic/Gulf coasts to Atlantic coast of South America and in the West Indies where there is only a small breeding population (*c.* 500 pairs): in the Bahamas, Hispaniola,

Cuba, Turks and Caicos Islands, Anguilla, Virgin Islands and Puerto Rico, but non-breeders are common year-round. Winters south through the breeding range to Argentina.
Status Common to locally common non-breeder; immatures and non-breeding adults present in all months. Birds in breeding plumage in May, absent in June, and juveniles from August.

Adult non-breeding has white trailing edge to outer primaries on underwing and white tail, November.

Adult breeding. Bill is entirely red without black distally unlike Caspian Tern, May.

First-winter has dark bar on secondaries, dark primaries on upperwing and dark on outer tail, December.

Sandwich Tern
Thalasseus sandvicensis

Adult non-breeding has black hindcrown and yellow tip to black bill, February.

Taxonomy Polytypic (3).
Description L 41–46cm (16–18in). Medium-sized crested tern. Appears white in flight, long slender black bill with yellow tip in most plumages, pale grey mantle, and black legs. Non-breeding adult has white forehead and forecrown, shaggy black hindcrown; in flight long pointed wings held at sharp angle, dark tips to outer primaries on upperwing, white tail moderately forked. Breeding adult (May–July) has crested black crown and pinkish wash on underparts. Juvenile has brown streaking on mantle and scapulars, rump and tail, bill may be all dark; first-winter has faint tip to bill, upperwing shows blackish secondaries, dark primaries and coverts, dark edges to tail. Cayenne Tern has bill yellowish or yellowish at base, legs partly yellow.
Similar species Gull-billed Tern has shorter, heavy black bill. Royal Tern has entirely orange or yellow bill.

Voice Grating s*kree-ik,* and high pitched agitated series.
Habitat and behaviour Barrier reefs, harbours, marine sounds and mangrove lagoons. Hovers and plunge dives from considerable height for flying fish.
Range Americas and Europe. *T. s. acuflavidus* breeds on the United States Atlantic-Gulf coast, cays off the Yucatan and Belize and the West Indies mainly in the Bahamas, Cuba, Hispaniola, Puerto Rico, Virgin Islands and Anguilla (Sombrero); winters in the West Indies to South America. The Cayenne Tern *T.s. eurygnatha* breeds in Netherlands Antilles, Venezuela, French Guiana to southern Argentina, and recently in Cuba, the United States and British Virgin Islands and may also occur in the Cayman Islands.
Status Uncommon passage migrant and winter visitor, September–April, the majority juvenile and first-winter birds.

First-winter has dark bar on secondaries, dark primaries and coverts on upperwing, and dark on outer tail.

Roseate Tern
Sterna dougallii

Adult is rare on passage, most often seen in breeding plumage, April to June.

Taxonomy Polytypic (5).
Description L 35–41cm (14–16in). Very long slender, pure white tern with long thin bill, pale grey mantle; adults have red legs. Non-breeding adult has rounded head, white forehead and forecrown, black hindcrown and black bill; in flight upperwing shows white inner edge to primaries and dark webs on three outer primaries, and white underparts deeply forked white tail with very long streamers (extend beyond folded wing at rest). Breeding adult has black crown and nape, underparts may be washed pink, bill black with red base briefly in peak breeding season. Juvenile has dark greyish crown, brown scalloping on back, dark carpal bar, black bill and legs, shorter tail.
Similar species Common Tern has less forked tail with dark sides and shorter tail streamers; juvenile has no scalloping on back. Non-breeding Forster's

Tern has oval black eye patch on white head; when breeding shows silvery-white primaries on upperwing.
Voice Usually silent; call *chi-vik*.
Habitat and behaviour Saline lagoons, marine sounds, inshore waters. Fast shallow wingbeats; makes long plunge dives for fish from high.
Range Cosmopolitan. *S. d. dougallii* breeds in North America on Atlantic-Gulf coast to Honduras, cays off Venezuela

and the West Indies. The main population is on the Puerto Rican Bank (US and British Virgin Islands and Puerto Rico), Bahamas, Turks and Caicos Islands, and Anguilla, colonies of over 60 pairs in Cuba, and Lesser Antilles; winters at sea. The Caribbean population has been designated as Near-Threatened by the United States Fish and Wildlife Service.
Status Rare short-stay passage migrant, April–June.

Juvenile has dark scallops on back, with prominent carpal bar and short tail, August. Cuba is the closest breeding colony.

Common Tern
Sterna hirundo

Taxonomy Polytypic (4).
Description L 33–40cm
(13–16in). Pale grey mantle,
white tail deeply forked
with blackish outer rectrices.
Non-breeding adult has white
forehead, black hindcrown and
nape joins black area behind
eye, white underparts, black
legs; in flight shows dark outer
primaries on trailing edge of
upperwing. Breeding adult
has black crown and nape,
grey wash to underparts, red
bill with black tip, red legs; in
flight shows grey underparts,
dark outer primaries on trailing
edge of underwing. Juvenile,
first-winter and non-breeding
have black carpal ba, white
underparts, black bill; juv.
(and some first-winter) have
red legs.

First-winter. Note black carpal bar and red legs, December.

Similar species Roseate whiter,
tail deeply forked with longer
streamers. Non-breeding
Forster's has oval black eye
patch on white head, breeding
shows silvery-white primaries
on upperwing.
Voice Harsh *kee-aarr.*
Habitat and behaviour
Harbours, sandy shores, and
saline lagoons. Plunge-dives for
fish and follows boats.
Range Holarctic. *S. h. hirundo*
breeds in North America and
locally to the West Indies;
winters mainly in South
America and West Indies.
Status Uncommon winter
visitor and passage migrant,
August–May.

Forster's Tern
Sterna forsteri

Taxonomy Monotypic.
Description L 35–42cm
(14–16.5in). White head with
oval black mask around and
behind eye is diagnostic in
non-breeding, first-winter and
juvenile. Non-breeding adult
has pale grey mantle, white
underparts, long, grey deeply
forked tail with white outer
rectrices, large black bill and
reddish legs. Breeding adult
has black cap from forehead
to nape, thick orange-red bill
with black tip and orange
legs; in flight shows silvery-
white primaries on upperwing.
Juvenile has ginger on crown
and back, reddish-black bill.
First-winter has speckled crown.
Similar species Only non-

First-winter has brownish-grey coverts and blackish on tertials, February.

breeding tern with black
ear-covert patch. Roseate Tern
has more deeply forked tail.
Common Tern non-breeding and
juvenile have black carpal bar.
Voice Silent.
Habitat and behaviour Saline
lagoons.
Range Breeds North America to
north-eastern Mexico; winters
to Panama and the West Indies,
rarely in the Bahamas, Greater
Antilles and Cayman Islands.
Status Rare passage migrant
and winter visitor, September–
April.

Least Tern
Sternula antillarum

Adult breeding plumage, Grand Cayman, May.

Local name Egg bird.
Taxonomy Polytypic (3).
Description L 22–24cm
(8.5–9.5in). Smallest white tern
with pale grey mantle, white
underparts. Non-breeding
adult has crown streaked
grey; black hind-crown, nape
and ear coverts, black bill
(in transition bill is blackish
proximally), greyish-yellow
legs, long narrow wings, short
tail slightly forked. Breeding
adult has white forehead, black
crown and nape joins black
eye-line, long yellow bill slightly
drooped at black tip, yellow
legs; in flight shows two black
primaries on leading edge of
wing. Juvenile has upperparts
scalloped brown; juvenile and
first-winter have thin blackish
post-ocular line and carpal bar.
Similar species All other white
terns are larger, none has yellow
bill with black tip and yellow
legs in breeding plumage.
Voice *Kre-ep* repeated, alarm
kit-kit-kit while diving on
intruders, and *zeet*.

Adult non-breeding has bill black distally, greyish-yellow legs and speckled crown, June.

Habitat and behaviour.
Flight swallow-like with fast
wingbeats; hovers with head
pointing downwards before
plunge-diving for fish in saline
and herbaceous wetlands
and inshore waters. Breeds
May–August; nest a scrape
on dredge spoil, lagoon
edge, Ironshore outcrops,
intermittently, on sand beaches
(previously the primary
habitat), 1–2 speckled eggs,
most young fledged by late
July. Nesting success low due
to predation and flooding
during May/June rains.
Range *S. a. antillarum* breeds
in North America to Honduras,

Adult breeding plumage. Two outer primaries are black, May.

Juvenile has dark scallops on back, black from eye to nape and dark carpal bar, June.

Venezuela, Netherlands
Antilles, and West Indies. Main
colonies are in the Bahamas,

Turks and Caicos Islands,
Greater Antilles, St Croix, Virgin
Islands (largest population),

Anguilla and small numbers
in the Cayman Islands and
Lesser Antilles; winters in South
America from the northern
Pacific coast east to Brazil,
where the majority occur. *S.
a. athalassos* and *S. a. browni*
breed in North America.
Status Fairly common summer
breeding visitor, April–early
October, on Grand Cayman. A
few pairs breed intermittently
on Little Cayman and Cayman
Brac. The population is
declining due to loss of habitat
and predation.

Fledgling camouflaged in nesting habitat, June.

Bridled Tern
Onychoprion anaethetus

Adult breeding has white forehead extending behind eye, Grand Cayman only, June.

Taxonomy Polytypic (6).

Description L 38cm (15in). Slender tern with dark grey mantle washed brownish, white underparts, black pointed bill and black legs. Non-breeding adult has greyish crown. Breeding adult has white forehead extending behind eye, black loreal line is same width from eye to bill, black crown and nape forming a 'bridle', narrow white collar; in flight shows narrow dark upperwing, grey tipped primaries on underwing, long tail dark and deeply forked with broad white edges. Juvenile has streaked crown, dark brownish-grey upperparts with wavy bands, and short tail.

Similar species Sooty Tern white on forehead does not extend behind eye, loreal line narrows towards bill. In flight underwing shows dark-tipped primaries.

Voice Call soft *weep*, harsher *krrrrr* over nest site.

Habitat and behaviour Marine sounds, cays and saline lagoons. The colony (*c.* 20 pairs) breeds on Vidal Cay, North Sound, May–July; nest in crevices and burrows under vegetation, one egg, young fledge after about two months. Buoyant graceful flight dipping for fish or squid from surface, seldom submerges in dive.

Range Pantropical. *O. a. recognita* is a Caribbean endemic subspecies breeding on islands off Belize and Venezuela, Netherlands Antilles and locally in the West Indies with the main colonies in the Bahamas, Turks and Caicos Islands, Greater Antilles (except Jamaica), Virgin Islands and Anguilla; ranges at sea in the Atlantic-Caribbean region.

Status Summer breeding visitor on Vidal Cay, off Grand Cayman, May–July; less than 20 pairs. Otherwise rare visitor, September–October, and March.

Juvenile has streaks on crown and pale wavy bands on back, July.

Sooty Tern
Onychoprion fuscatus

Taxonomy Polytypic (8).
Description L 38–43cm
(15–17in). Large black and
white tern. Non-breeding adult
has crown and nape greyish,
blackish upperparts, black bill
and legs. Breeding adult has
white on forehead extending
only to eye, black loreal line
narrows towards bill, black
crown, nape and upperparts. In
flight has black, deeply forked
tail with narrow white outer
edge, primaries on underwing
blackish with white coverts.
Juvenile is black, spotted with
white on mantle and coverts.
Similar species Bridled Tern
has white forehead extending
behind eye, only tips of
primaries are blackish on
underwing.
Voice Usually silent, call w*ide-a-
wake.*
Habitat and behaviour
Fringing reefs, coastal waters;
makes shallow plunge dives or
lifts fish or squid from surface

Adult breeding. White on forehead extends to eye in winter and summer
plumages, June.

in deep water.
Range Pantropical. *O. f.
fuscatus* breeds in the Atlantic-
Gulf-Caribbean of North
America (irregularly), Mexico,
cays off Venezuela and in the
West Indies: where the largest
colonies are in the Bahamas,
Turks and Caicos Islands,
Greater Antilles, Anguilla and St

Lucia. Highly pelagic, thought
to sleep on the wing, staying
aloft continually while ranging
at sea in the Atlantic-Caribbean
and tropical Atlantic to West
Africa. The most abundant tern
in the Caribbean estimated over
300,000 pairs.
Status Rare visitor, April–
September.

Black Tern
Chlidonias niger

Juvenile/first-winter has black ear-coverts and brownish patch on sides of breast, May.

Taxonomy Polytypic (2).
Description L 23–26cm (9–10in). Small dark tern with small black bill, blackish patch extending onto side of breast in all plumages, short notched grey tail, long wings extend beyond tail at rest; in flight upperwing and underwing are grey. Non-breeding adult has blackish hindcrown joining round black ear coverts, dark grey upperparts, and blackish legs. Breeding adult black except for dark grey mantle and wings, white undertail-coverts and black legs. Juvenile has streaked crown, white scaling on brownish back, pinkish-red legs.
Similar species Only small tern with black ear-coverts joining blackish hindcrown and distinctive dark patch on sides of breast. Non-breeding Forster's Tern has white head and oval black patch over eye, pale grey mantle and red legs. Least Tern has yellow legs and deeply forked tail in all plumages.
Voice Sharp *kyck*, usually silent.
Habitat and behaviour Herbaceous wetlands, ponds and flooded marl pits on Grand Cayman, also saline lagoons.
Buoyant flight with slow deep wingbeat on broad wings (like a nighthawk), frequently dips fish and insects from surface and may dive; perches on posts and roosts with waders.
Range Breeds North America and Eurasia. *C. n. surinamensis* breeds in North America; winters from Panama to both coasts of South America from Peru to Surinam. Rare and irregular on passage throughout the West Indies.
Status Uncommon passage migrant, August–November and February–June. One January record after a storm.

Black Skimmer
Rhynchops niger

Adult non-breeding has whitish nape and brown on mantle; bill is diagnostic, October.

Taxonomy Polytypic (3).
Description L 40–51cm (16–20in). Unmistakable. Large black-tipped, laterally compressed, red bill, has longer, thinner lower mandible projecting beyond upper mandible; black mantle, very long pointed wings, buoyant flight and head held low. Non-breeding adult has black cap, white forehead, nuchal collar and underparts. Breeding adult has black crown and nape. Juvenile has brownish streaked crown and upperparts, and blackish red bill.
Similar species None.
Voice Silent.
Habitat and behaviour Marine sounds, lagoons. Flies low over water with bill open and lower mandible submerged, cutting water's surface for small fish.
Range *R. n. niger* breeds in North America on United States Pacific and Atlantic-Gulf coasts and Mexico; winters from Florida to Panama and Brazil, and casually in the Bahamas, Greater Antilles and Cayman Islands. Two additional subspecies breed in South America.
Status Rare winter visitor and passage migrant, August–May.

Rock Pigeon
Columba livia

Taxonomy Polytypic (13). Introduced.

Description L 33–36cm (13–14in). Individuals show wide colour variation (brown, white, multi-coloured and dark), the most common has a dark bluish-grey head and nape, iridescent greenish-purple sheen on sides of neck, pale grey back and wing-coverts with two black bands on closed wing, orange iris, white cere, black bill, and legs reddish-plum. In flight shows white rump and black terminal band on tail.

Similar species None.

Voice Call a low series of cooing notes.

Habitat and behaviour Forages for seeds on the ground in urban areas and rubbish tips. Breeds throughout the year in a rough stick nest on ledges of low buildings, clutch of 2 white eggs.

Rock Pigeon is now established on Grand Cayman and Cayman Brac.

Range Introduced and established worldwide including the West Indies and the Cayman Islands.

Status Introduced; feral populations resident and breeding on Grand Cayman since 1995, and Cayman Brac since 1997; vagrant to Little Cayman.

White-crowned Pigeon
Patagioenas leucocephala

Local name Bald Pate.

Taxonomy Monotypic.

Description L 33–36cm (13–14in). Large, entirely charcoal-grey pigeon with white crown, violet and green iridescence on nape and sides of neck giving a scaly effect; iris white, orbital skin pinkish-white, bill red tipped with grey and legs red. Male has immaculate white crown from forehead to hind-crown; female has crown greyish-white, less extensive. Juvenile is entirely dark brown with back and wing coverts edged buff; immature has forecrown greyish and dull

Breeding male with white crown, iridescence on neck, pale iris and grey tip to bill, April.

Female with grey crown, November.

brownish-grey plumage, iris dark: adult plumage develops in second year.

Similar species None.

Voice Deep throaty *croo cru* (rising) *cu-cruuu,* and soft rolling *cru-cruu.*

Habitat and behaviour Frugivorous arboreal forager in all habitats, taking fruits of Sea-grape, Royal Palm, Red Birch, Silver Thatch, Cabbage Tree, Pepper Cinnamon, Wild Fig and Black Mangrove, and Jasmine flowers. Breeds on Grand Cayman, March–September, and on Little Cayman and Cayman Brac, May–September, with mangrove forest and shrubland preferred by larger colonies; pairs and small groups in dry forest and shrubland (formerly in littoral Sea-grape), and single pairs in urban areas. Builds rough twig nest 4–20m (13–62ft) above ground with 1–2 white eggs, young are dependent on parents for up to 40 days.

Range Caribbean species, breeding in the Florida Keys, islands in the western Caribbean off Yucatan Peninsula, Belize, Honduras, Panama, and in the West Indies in the Bahamas, Greater Antilles, Cayman Islands and Lesser Antilles (rare south of Antigua); in decline throughout most of its range. Wanders widely between the Antillean islands.

Status Most are migratory, arriving in late January–February, becoming fairly common to common, April–September, and rare to absent, November–January. Up to the mid-1990s, an abundant year round resident; it remains a game species and is hunted when young are still in the nest, contributing to its decline together with loss of habitat.

Male feeding on Red Birch, a staple for landbirds as trees bear fruit throughout the year, April.

White-winged Dove
Zenaida asiatica

Adult male with black streak below eye, blue orbital ring, orange iris and white band on closed wing, April.

Local name White Wing.
Taxonomy Polytypic (3).
Description L 28–30cm (11–12in). Adult plumage brownish-cinnamon, iridescent violet on crown, nape and hindneck, black streak below pale ear coverts, wide blue orbital ring and orange iris, whitish throat, greyish abdomen and rump, bill black and feet red. White band on closed wing shows as white wing-coverts on black wings in flight, square tail with black subterminal band and wide white terminal band. Juvenile pale and greyish, iridescence absent, iris dark.
Similar species Zenaida Dove is smaller with white spot on closed wing that shows as a narrow white band on trailing edge of secondaries.
Voice *Cru-cu-ca-roo* in level pitch, a long series *cura-caa-cura-caa*, and call *cura-croo*.
Habitat and behaviour Fast direct flight. Forages on the ground for seeds and in trees for fruits in all terrestrial habitats, especially urban areas and disturbed dry shrubland and second growth. Aggressive, raising one wing in threat display and attacking by wing-buffeting. Breeds year-round on Grand Cayman, and mainly April–August on Little Cayman and Cayman Brac. Single pairs and small flocks found in all habitats, larger colonies in mangrove forest on Grand Cayman, and dry shrubland and Buttonwood on Little Cayman. Sparse twig nest, placed 1.3–20m (4–60ft) up; 2 white eggs and two to three broods usual on Grand Cayman.

Flocks are common in urban and disturbed areas.

Range Americas. *Z. a. asiatica* breeds from Texas through Middle America and the West Indies in the Bahamas, Greater Antilles (it colonised Cuba in 1923, Isle of Youth in 1959, Puerto Rico in 1943), Cayman Islands in 1935, San Andres and Providencia in the western Caribbean. Northern birds migrate south in winter.

Status Majority are migrants breeding on the three islands, fairly common-common, February–October, becoming uncommon, November–January. Populations fluctuate and crashed following three hurricanes during 2004–09, but recovering.

Adult exhibiting white band on upperwings, grey underwing-coverts and white terminal tail-band, July.

Zenaida Dove
Zenaida aurita

Local name Pea Dove, Big Dove.

Taxonomy Polytypic (3).

Description L 25–28cm (10–11in). Male has golden wash on cinnamon-brown head, iridescent violet on hindneck, six to eight black spots on wing-coverts, vinaceous-brown underparts, violet blue mark above and below ear coverts, blue-orbital ring and dark iris. Tail rounded with black subterminal band and light grey tips to outer rectrices, legs red. White spot on closed wing shows as white band on trailing edge of secondaries in flight. Female similar but duller. Juvenile has buff-white edges to back and wing-coverts and lacks iridescence.

Similar species White-winged Dove is larger with broad white band along closed wing.

Voice Echoing *ku-ra* (rising) *ku*

Adult has iridescence on neck, black spots on coverts and white spot on closed wing, February.

coo coo ('Moses preach God's Word') followed by sighing *ooh ah ah* or *croo*.

Habitat and behaviour Forages mainly on the ground for seeds, also fruits in trees, in dry forest

Adult displaying white trailing edge to secondaries, dark primaries and light grey tips to outer rectrices; note violet marks above and below ear-coverts, December.

and shrubland, littoral areas and shores on Little Cayman, and agricultural land and second growth. Sunbathes and rain bathes lying prone on the ground with one wing outstretched. Breeds as single pairs and in small loose colonies, February–October on Grand Cayman, and April–July on Little Cayman and Cayman Brac, depending on rainfall. Pair bonds usually retained and territories defended; nest of rough sticks at low to middle levels, often in trees with bromeliads or in Silver Thatch Palms; 2 white eggs, only one young may be raised, but several broods usual in optimum conditions.

Range Resident on the northern Yucatan coast and islands offshore, and the West Indies. *Z. a. zenaida* is resident in the Bahamas, Greater Antilles to the Virgin Islands, and Cayman Islands. *Z. a. aurita* breeds in the Lesser Antilles.

Status Breeding resident, fairly common in the east, north and central districts of Grand Cayman and rare west of Savannah due to habitat loss; numbers are adversely affected by the White-wing Dove populations; fairly common to common on Little Cayman and Cayman Brac where it is the dominant large dove.

Pair nesting in Red Birch tree, April.

Mourning Dove
Zenaida macroura

Taxonomy Polytypic (5).
Description L 28–33cm
(11–13in). Similar to Zenaida
Dove in plumage except for
diagnostic long wedge-shaped
tail with white edges, and
unmarked wings in flight. Male
has purplish iridescence on
hindneck and sides of neck, and
small streak below eye. Female
paler with less iridescence.
Juvenile is grey-brown with back
and wing-coverts edged buffy,
buffy ear-coverts and face and
faint spotting on breast.
Similar species Only escaped
very pale Eurasian Collared-
Dove has a long tail.
Voice Silent.
Habitat and behaviour
Disturbed and urban/littoral
areas.
Range Breeds in North
America, Middle America
and in the West Indies: in the

Adult shows similarity to Zenaida Dove except for long tail, March.

Bahamas and Greater Antilles,
where the subspecies *Z. m.
macroura* is resident. North
American birds winter in the
breeding range.
Status Rare passage migrant
mainly in spring. There is one
January record.

Common Ground Dove
Columbina passerina

Taxonomy Polytypic (18).
Description 15–18cm
(5.75–7in). Male has bluish-
grey crown and upperparts,
dark blue iridescent spots on
cinnamon-brown wing-coverts
and tertials, pinkish-cinnamon
forehead, face and throat;
beautifully marked on breast
and sides of neck with pinkish
pale grey feathers darkly
edged, giving a scaly effect;
bill pink with black tip, legs
pink. Female is slightly duller
and crown is sandy-grey;
juvenile similar but back and
wing-coverts edged whitish,
scaly head and greyish breast,
bill dark. Adults in flight

Adult male has scaling effect on head and breast and iridescent blue spots on
wing-coverts, December.

Adult female has brown or copper spots on wings, December.

show rufous primaries and underwings and black tail with white outer edges.

Similar species Only diminutive dove.

Voice Call steady, level single *coo coo coo coo* or *co-coo co-coo co-coo* and whooping *hoap-hoap-hoap*.

Habitat and behaviour Usually in pairs foraging on the ground for seeds in open dry shrubland, littoral areas to the sea edge, rough pasture, urban areas and gardens (at bird feeders and around hotels) and plantations. Bathes in the sun and light rain with one outstretched wing on the ground. On Grand Cayman breeds throughout the year, mainly March–August, and on Little Cayman and Cayman Brac, January–September. Pair bonds are retained throughout the year. Nest is a scrape on the ground lined with grasses, or flimsy mat of grass and fibre at low to mid levels, often in Silver Thatch or Coconut Palms or Sea-grape, and on window ledges; two white eggs laid, young fed by both parents and remain as family group.

Range Breeds in the southern United States through Middle America to Ecuador and Brazil, islands in the western Caribbean and throughout the West Indies. *C. p. insularis* occurs in Cuba, Cayman Islands and Hispaniola and associated Islands.

Status Common to abundant resident.

Adult showing rufous primaries and underwing-coverts, April.

Juvenile has back and wing-coverts edged pale and all-dark bill, July.

Caribbean Dove
Leptotila jamaicensis

Male shows iridescence from hindneck to nape and white iris, Grand Cayman only, April.

Local name White Belly.
Taxonomy Polytypic (4).
Description L 30–33cm
(12–13in). Stocky dove with
grey crown; ivory-white
forehead, face and throat. Male
has violet-pink and bronze
iridescence on nape and upper
back, rest of upperparts olive-
brown, iris white with purple
orbital skin, iridescent greenish
sheen on rosy-brown sides of
neck and breast, separated by a
narrow white notch from folded
wing, breast greyish-white;
abdomen and undertail-coverts
white; bill black, legs and feet
red and fleshy. Female has less
iridescence. Juvenile has duller
plumage with rufous edges to
back and wing-coverts, sides
of breast barred brownish. In
flight shows brown primaries
edged with white, chestnut
underwings and central tail
greyish-brown with outer
rectrices black with white tips.
Similar species Only dove with
white underparts.

Voice Long mournful *cru cru
cru cru-aaaaa* descending
and repeated, and duets with
bonded pairs in the breeding
season.

Habitat and behaviour Solitary
or in pairs; forages for seeds on
the ground, with head moving
back and forth and tail up
and down, in dry forest and

Adult female has brown or copper spots on wings, December.

shrubland and adjoining mangrove edge of Central Mangrove Wetland and Buttonwood wetlands in the east, shaded plantations and gardens in the central and eastern districts; it is absent from coastal areas and the interior of mangrove. Regularly observed walking on the pinnacled limestone in the Mastic Reserve. It breeds in dry forest and woodland, March–September, building stick nests from low to high levels often in trees with large Bromeliads or Banana Orchids; two white eggs laid, both adults brood and raise young.
Range Resident in Middle America (subspecies *L. j. gaumeri* in northern Yucatan Peninsula and islands off the east coast of Mexico, Honduras and Belize), *L. j. neoxena* on San Andres, *L. j. jamaicensis* on Jamaica, and *L. j. collaris* on Grand Cayman. Introduced on New Providence, Bahamas about 1930.
Status Uncommon resident endemic subspecies *L. j. collaris* on Grand Cayman only, at North Side and the central and eastern districts.

Adult on nest in dead Red Birch tree, September.

Monk Parakeet
Myiopsitta monachus

Taxonomy Polytypic (3).
Description L 29cms (11.5in). Adult has pale grey forehead, lores, cheeks and throat; bright green nape, upperparts and very long tail; breast pale grey with brownish feathers giving a scaly effect, olive-yellow band across upper abdomen, greenish-yellow lower abdomen and undertail-coverts. In flight shows dark blue primaries and secondaries on upperwing, appears blue-black on underwing. Juvenile similar, with forehead and cheeks greenish.
Similar species Only parakeet.
Voice Loud continuous sharp *skreet*.
Habitat and behaviour Urban areas with trees; noisy and gregarious, roosting and breeding in large communal interwoven stick nests with multiple entrances, attached to a palm tree or utility pole; breeds in fall and winter.
Range South America. Introduced in North America and the West Indies: in the Bahamas (Eleuthera), Cayman Islands, Puerto Rico, Virgin Islands, and Guadeloupe.
Status Introduced accidentally. Small feral breeding population, from escaped cage birds, was established on Grand Cayman *c.* 1992; the expanding population was greatly reduced by Hurricane Ivan in 2004.

Adult. Feral breeder that builds large communal nests in urban areas, Grand Cayman, November.

Cuban Parrot (Cayman Parrot)
Amazona leucocephala

Local names Parrot, Brac
Parrot.
Alternative English names
Rose-throated Parrot (Raffaele
et al 1998) and Cuban Amazon
(del Hoyo *et al.* 1997).
Taxonomy Polytypic (4).
Description L 28–33cm
(11–12in). Short-tailed amazon
with iridescent green plumage
edged blackish, wide whitish
eye-ring and orbital skin,
blackish ear-coverts, thick
pale bill with deeply curved
culmen, pale legs and feet.
Blue band on wing shows as
blue primaries and secondaries
in flight, undertail-coverts
yellowish-green, tail green
with blue outer edges and
red tips to uppertail-coverts.

Adult male, Grand Cayman subspecies, has red cheeks continuous with throat, and some pink feathers on forecrown, January.

Adult male, Cayman Brac subspecies, with deep maroon area on abdomen, February.

Grand Cayman adult (*A. l. caymanensis*) has white forecrown with varying amounts of red feathers, rose-red cheeks, chin and throat; black ear coverts, and varied amounts of reddish-pink on lower abdomen; male is larger and brighter with red of cheek patch and throat almost continuous. Cayman Brac adult (*A. l. hesterna*) is smaller with darker green feathers edged black; forecrown usually entirely white and extends further into mid-crown but occasionally edged with a few pink feathers; upper-tail and undertail-coverts yellowish tipped with red, extensive maroon areas on abdomen of male, smaller in female. Juvenile is smaller, bright leaf-green overall with yellow feathers on forecrown, and yellow and pink feathers on cheeks and throat; on

Cayman Brac immatures are occasionally seen with entirely green forecrown.
Similar species Yellow-crowned Parrot on Grand Cayman is green overall with yellow on crown or nape.
Voice Detected usually by calls. Very vocal with unmusical whistles, bugles and squawks, especially when preparing to leave the roost. Wide range of calls for flight, contact, territory and threat; flight call of *hesterna* is a hoarser *d-de* higher pitched than *caymanansis*, also has louder and quicker-paced *ke ke ke* chatter between pairs. Both have a conversational murmur among feeding groups.
Habitat and behaviour
Appears black in flight unless in sunlight, with blunt head and fast downward wingbeat with wings not

raised above horizontal. Fly in pairs except when female on the nest, and bond for life. Both subspecies forage in all terrestrial habitats, including gardens and plantations, for fruits and seeds, young leaves and flowers, frequently using tongue and feet; preferred food species include Red Birch, Wild Fig, Sea Grape, Silver Thatch, CoCo Plum, fruit trees in plantations e.g. Mango and Papaya, and vines e.g. Smilax. *A. l. caymanensis* also forages in Black and White Mangrove Forest, Logwood/Buttonwood, littoral areas including low (<1 m) beach ridge shrubs; easily observed on the Mastic Trail and Botanic Park, usually absent from Barkers and the interior of the Central Mangrove Wetland and mangrove shrubland. *A. l. hesterna* occurs as a small population on Cayman Brac, foraging on the bluff and on both coastal plains, but chiefly in urban areas with fruit trees on the north coast; absent from the extreme eastern bluff and wetlands; individuals and small flocks fly occasionally to forage in eastern Little Cayman during the day. This subspecies

Pair, Grand Cayman subspecies, beside nest cavity, March.

is silent, secretive and hard to observe. The best sightings are during January–April and August–December on the bluff on the main Major Donald Drive at the National Trust Parrot Reserve car park just after dawn and just before dusk, as

the parrots fly across the road from or to their roosts; also at Centennial Park if fruit trees are ripe, at the Aston Rutty centre and on the north coast road just east of Stake Bay in the late afternoon. Permanent pair bonds reinforced by courtship and nest preparation; breeds, March–July, nests in deep cavities in mature forest trees; clutch 1–5 (usually 1–4) round white eggs; female only incubates for *c.* 28 days and both parents feed; 1–3 young fledge after *c.* 60 days and beg from parents for several weeks, remaining in small family groups, learning food types and survival strategies. In Grand Cayman the majority breed east of Savannah in dry forest, Black Mangrove Forest in the Central Mangrove Wetland, often in old woodpecker holes in Royal and Silver Thatch palms, snags, and

Adult, Grand Cayman subspecies. Feathers on back and coverts have dark edging reduced compared to Brac subspecies, February.

Cayman Brac subspecies showing white forecrown, March.

Cayman Brac subspecies in bluff forest showing blue primaries and secondaries, April.

large single trees in urban areas, gardens and rough pasture. In Cayman Brac breeding is confined to mature dry forest on the bluff, with Cedar preferred: there the absence of resident woodpeckers is a major constraint and results in requirement for dead or partially dead trees for nest cavities.

Range West Indian endemic.

Resident on Cuba, Bahamas and Cayman Islands. Endemic subspecies A. l. leucocephala in Cuba and Isle of Youth, A. l. bahamensis in the Bahamas (Abaco and Great Inagua), and two subspecies on the Cayman Islands.

Status Endemic subspecies A. l. caymanensis is fairly common to locally common on Grand Cayman. A. l. hesterna occurs as a small population on Cayman Brac, locally common when flocking in pre-breeding season, otherwise more often seen than heard; it occasionally visits Little Cayman, where it was reported breeding up

to 1940 but not since the 1944 hurricane. The parrot is a protected species and some forest breeding habitat is protected in the Mastic Reserve, the Salina and Botanic Park on Grand Cayman and the Parrot Reserve on Cayman Brac. The most serious threat is fragmentation and clearing of dry and mangrove forest, which removes mature nesting trees. Other threats are illegal shooting and trapping for the pet trade (when nesting trees are destroyed by cutting into the bole to remove young) and feral cats. Conservation Status: Cites 1; Near-threatened.

Immature, Cayman Brac subspecies. Note dull plumage and absence of white forecrown and red of cheeks and throat, April.

Juvenile, Grand Cayman subspecies, has yellow feathers on forehead, July.

Yellow-crowned Parrot
Amazona ochrocephala

Adult shows yellow on forehead and nape, and red on bend of wing; population confined to south-central Grand Cayman, June.

Taxonomy Polytypic (10). At least two subspecies present.
Description L 35–38cm (14–15in). Single bird is yellow-naped (likely *A. o. auropalliata*); a pair have yellow forehead and yellow on nape (likely *A. o. parvipes*). Green overall with underparts lighter, face and cheeks bright green, white eye-ring, red on bend of wing, green tail with yellowish top and red on outer feathers. In flight shows red secondaries and flight feathers edged dark blue. Juvenile green; immature duller with less yellow on nape.

Similar species Cuban Parrot has no yellow on crown or nape.
Voice Higher pitched and faster than Cuban Parrot, with shorter bugles.
Habitat and behaviour Woodland, mangrove and gardens, foraging on cultivated fruit trees, exotic flowering trees and bananas. Breeds in tree cavities, April–July, 1–2 young observed.
Range *A. ochrocephala* is resident in Middle America, Panama, Colombia, Venezuela, Trinidad, the Guianas to northern Brazil. The ten subspecies form three species groups.
Status Status uncertain, probably breeding. This species (as two, perhaps three, subspecies) was introduced accidentally in 1990s and formed a mixed flock. Confirmed breeding 1993–2003. It was thought to have perished in the 2004 hurricane but 2009–2012 one individual (yellow-nape) and a pair (yellow on forehead and nape) photographed in the Savannah-North Sounds Estates area, Grand Cayman.

Yellow-billed Cuckoo
Coccyzus americanus

Taxonomy Monotypic.
Description L 28–32cm
(11–12.5in). Long slender bird
with greyish-brown upperparts,
grey orbital ring; long, stout
slightly down-curved bill
almost entirely yellow except
for dark central culmen and
tip, underparts white, long tail
with broad white markings
seen from underneath. In flight
shows rufous primaries with
black trailing edge and large
white tail spots on edges of
long tail. Juvenile similar but
orbital ring yellow.
Similar species Mangrove
Cuckoo has black mask, buffy
underparts and brown wings
show no rufous in flight. Black-
billed Cuckoo (vagrant) has grey
bill and small tail spots.
Voice Silent.
Habitat and behaviour Littoral
areas, dry and mangrove
shrubland and rough pasture.
Range Breeds in the United
States and Mexico and,
uncommonly, as a summer
visitor in the Greater and Lesser
Antilles. Winters in South
America, and throughout the

First-year on passage. Note yellow iris, April.

breeding range; on passage
in the Bahamas, Cuba,
Hispaniola, Jamaica and
Cayman Islands.

Status Uncommon to locally
common passage migrant,
August–November and April–
June; one January record.

Mangrove Cuckoo
Coccyzus minor

Local name Rain bird.
Taxonomy Monotypic.
Description L 28–30cm
(11–12in). Long slender bird,
with greyish-brown crown and
hazel-brown upperparts, black
band from lores over eye to
ear coverts, yellow orbital ring;
long heavy down-curved bill
with most of lower mandible

yellow with dark tip; white
throat, breast pale buffy-
cinnamon becoming richer on
abdomen and undertail-coverts,
very long tail with large white
spots on black underside. In
gliding flight shows unmarked
wings (no rufous) and tail with
black and white edges. Juvenile
lacks black on face, eye-ring

grey, back and wing-coverts
edged cinnamon, underparts
buffy-cinnamon and tail spots
less distinct.
Similar species Yellow-billed
and Black-billed Cuckoos have
white underparts and lack black
ear coverts; the former has
rufous primaries.
Voice Long nasal series

Adult has cinnamon underparts and dark upper mandible, March.

ge-ge-ge... gou-gou-goul-goul, slower at end.

Habitat and behaviour Inactive and approachable; widely distributed, foraging for insects and spiders slowly at mid to low elevations in dry shrubland and mangrove/dry forest edge on Grand Cayman, but usually absent from mangrove. Heard more often than seen, calling before the onset of rain. Builds sparse twig nest platform often below 3m up in dry forest understorey, woodland or shrubland; lays 2–3 eggs, April–June.

Range Breeds in southern Florida, Middle America to northern South America, and the West Indies; winters throughout the breeding range.

Status Fairly common breeding resident.

Adult, note long tail, November.

Smooth-billed Ani
Crotophaga ani

Local Name Black Arnold, Old Arnold.
Taxonomy Monotypic.
Description L 30–33cm (12–13in). A member of the cuckoo family, often reported as a black parrot. Adult has blue-black plumage with iridescent bronze sheen on back and wings; head and neck feathers are short and scaly with bare skin around eye; large parrot-like black bill with ridged curved culmen projecting to crown, very long tail. Juvenile is smaller with brownish-black plumage edged grey.
Similar species None, heavy bill is diagnostic.
Voice Rising whistled *keuu-iik*, and repeated loudly, low *ann-ee*, also whistles and harsh squawks.
Habitat and behaviour Small flocks usual in disturbed habitats with grassland, roadsides and gardens where forages for arthropods; also in trees, taking young birds, lizards, snakes and fruits in all habitats including the forest canopy. Flap-and-glide laboured flight on short rounded wings; very reptilian

Adult has keel-like ridge on curved culmen, October.

aspect, tumbling on landing with wings spread and ragged plumage. Breeds throughout the year on Grand Cayman and from spring to fall on Little Cayman and Cayman Brac. Bulky communal nests are shared by several females; 4–5 bluish eggs with flaky white coating laid in layers separated by leaves; only the top layer hatches.
Range Breeds in Florida, Middle America and islands off the coast, Panama, South America and the West Indies.
Status Common breeding resident, and increasing as disturbed habitat increases.

Juveniles, March.

Barn Owl
Tyto alba

Local name Screech Owl, White Owl.

Taxonomy Polytypic (28).

Description L 29–43cm (12–17in). Sexes similar but female larger and darker. Adult has large rounded head, heart-shaped white facial disc with large dark eyes, body tapers to short tail, long wings and feathered legs with well-developed talons. Upperparts flecked orange and gold with greyish spots, underparts and underwing whitish with flecks of brown; appears entirely white at night, with plumage and long broad wings adapted for silent flight, hence its 'ghostly' image.

Adult in flight, April.

Juvenile has darker brown upperparts and face and is more spotted and streaked.

Similar species None.

Voice Hisses, squeaks and clicks, a shriek *shreeeeeeeee* heard while hunting, 2-syllable screech during breeding displays and variety of noises at the nest.

Habitat and behaviour In all habitats with suitable nesting and roosting sites: cavities in large trees in mangrove and dry forest, caves high on the bluff and at ground level, single trees in gardens, lofts of houses, buildings under construction, airport terminal, sports stadia and abandoned boats, buildings and cars. Mainly nocturnal and crepuscular hunter, also observed in mid-morning or mid-afternoon on dark cloudy days; hunts mainly birds (comprise 60% of pellets), also lizards, rats and mice and bats. Breeds in all months, peaks December–March, preceded by noisy pair-bonding displays. Highly territorial at nest site and pairs mate for life, up to seven round white eggs laid on bare substrate, incubated by female which is fed by the male. Young fledge after about 10–11 weeks often remaining with the parents for several months. Second brood is raised in optimum years.

Range Cosmopolitan. Breeds throughout the Americas and West Indies. *T. a. furcata* is confined to Cuba, Jamaica and Cayman Islands.

Status Uncommon breeding resident. Populations on Grand Cayman and Cayman Brac have declined due to hurricanes, habitat loss, poisoning by rodenticides and road kills (especially juveniles).

Adult female, April.

Juvenile in abandoned house. April.

Short-eared Owl
Asio flammeus

Adult. Note black tips to primaries and black carpal patch on wing.

Taxonomy Polytypic (10).
Description L 35–43cm (14–17in). Head large with round, brownish facial disc, black patches around large yellow iris; plumage cryptic with dark brown upperparts spotted buff, breast buff with heavy streaks, tail barred with white terminal band, short black bill, feet and legs pale. Female larger, darker and more heavily streaked. Juvenile has dark face, browner upperparts edged buff and fainter streaked underparts. Buoyant flight on long, broad pointed wings, upperwing shows large yellowish-buff patch at base of outer primaries contrasting with black carpal patches and black tips to primaries.
Similar species None.
Voice Hissing, wing clap and bill snapping. Call *beee-ow*; and *kuk-kuk-kuk*.
Habitat and behaviour Grassland and disturbed areas: rough pasture, land cleared for developments, golf courses, lagoon edges and wet meadows. Active hunter during the day, night, dawn and dusk, quartering slowly over land for lizards, rats, bats, birds and insects detected by sight and sound; also takes road kills. Aerial courtship display observed. Roosts and breeds on the ground in grassland, where two nests lined with grasses and feathers had two and five white eggs respectively: on Grand Cayman and Cayman Brac, December and April.
Range Holarctic. Breeds in North and South America, and western Greater Antilles.

Northern North American subspecies *A. f. flammeus* winters to southern United States, Mexico and Greater Antilles. In the 1990s, *A. f. domingensis* underwent a population explosion in Cuba and Hispaniola and is now breeding and resident in the Cayman Islands.
Status Uncommon and irregular breeding resident on Grand Cayman and Cayman Brac; no breeding records on Little Cayman, where it is a migrant. Known nomadic wanderer, colonisation probably related to post-breeding dispersal of the Cuban subspecies (six injured juveniles were treated by the National Trust, 1996–1999). Migrants may also come from North America.

Antillean Nighthawk
Chordeiles gundlachii

Adult male showing white throat-patch and tail-spots, July.

Local Name Rickery-dick.
Taxonomy Polytypic (2).
Description L 20–25cm (8–10in). Upperparts blackish-brown speckled whitish-grey, cinnamon, russet, black and tawny; breast white, rest of underparts buffy, all barred greyish, large flat head, large eyes, whitish supercilium and V on scapulars, small bill with a wide gape, and small weak legs. Wings do not extend beyond tail at rest, and in flight appear long, slender and pointed, bent in a V at the carpal joint; blackish primaries. Male has white throat patch, in flight shows broad white cross-band on five outer primaries and broad white sub-terminal tail band. Female has throat patch smaller, less distinct white band on

Adult female, May.

Male has white subterminal tail-band, lacking in female, June.

Male fanning tail during courtship, July.

primaries, throat buffy, white subterminal tail band absent. Juvenile similar to female but underparts heavily barred, white trailing edge to wing, pale panel on the upper wing, primaries and secondaries tipped with white.
Similar species Common Nighthawk is virtually identical; larger and wings extend beyond tail at rest but separation in the field relies on its single syllable call.
Voice Constant 3–5 syllable *rickery-dick* is diagnostic.
Habitat and behaviour Diurnal and nocturnal insectivore foraging singly, in pairs or large flocks, with extreme aerial agility, over all open disturbed areas, sometimes at high elevations. Perches lengthways on tree branches. Breeds, May–August, male courtship includes 'booming' in a series of steep dives. Nest a scrape on leafy or stony ground, in rock clefts, marl pits, spoil banks, the Ironshore, lagoon edges, and open woodland, clutch 1–2, eggs grey mottled with olive, incubated by female.
Range *C. g. gundlachii* breeds in the Bahamas, Greater Antilles, Cayman Islands and southern Florida, it is thought to winter in South America.
Status Summer breeding visitor, March–September, with a few in October. Locally abundant in some years. Numbers declining due to loss of habitat and increased predation.

Eggs, mottled with olive, July.

Chuck-will's-widow
Caprimulgus carolinensis

Taxonomy Monotypic.
Description L 31 cm (12 in).
Adult upperparts are rich
brown, buff and grey, streaked
blackish-brown; wing-coverts
brown heavily spotted with
white and buff, scapulars
spotted blackish, whitish band
around throat, rictal bristles,
underparts brownish, spotted
tawny and buff. Dark wings
lack cross-bands. Male larger,
with white inner webs on three
outer tail feathers, absent in
female.
Similar species Antillean
Nighthawk has white bands on
primaries in flight.
Voice Silent.
Habitat and behaviour Urban
littoral woodland and forest
edge. Nocturnal, feeding
on insects from a perch or
hawking. Roosts during the day
in trees from which not easily
disturbed.
Range. Breeds in eastern

Adult perched.

North America and winters
in southern United States,
Middle America to northern
South America and Bahamas
and Greater Antilles (common
in Hispaniola and Cuba,
uncommon to rare elsewhere);
on passage in the Cayman
Islands, vagrant to northern
Lesser Antilles.
Status Rare passage migrant,
September–December and
February–May.

Adult, note cryptic plumage and
wide mouth.

Black Swift
Cypseloides niger

Taxonomy Polytypic (3).
Description L 15–18 cm (5.75–7in). Adult is large and
appears black overall with pale grey forecrown, very long
wings forming an arc in flight, and broad, short tail,
slightly forked in male.
Similar species Chimney Swift is smaller with shorter
square tail, grey brown plumage, and wing shape does
not form a wide arc, flight is faster with rapid wingbeats.
Voice Silent.
Habitat and behaviour Hawks insects, wings held below
horizontal in glide; observed only in rapidly moving
migration flights usually with swallows.
Range Breeds in western North America over a scattered
range, Middle America and the West Indies, where *C. n.
niger* is a common resident on Jamaica and Hispaniola,
local on Cuba and a summer breeder in the rest of
the West Indies, excluding the Bahamas and Cayman
Islands. Little is known of its migration pattern although
North American and West Indies subspecies winter in
South America.
Status Rare passage migrant, usually single birds with
swallows, October and May.

First-winter immature in flight. Note white tips
to feathers on underparts.

Chimney Swift
Chaetura pelagica

Taxonomy Monotypic.
Description L 13cm (5in). Small swift, with
large head, dark grey-brown upperparts, darker
rump, paler underparts with pale buffy throat,
long wings with bent outer wing; and very short
rounded tail with protruding spines (barely
visible).
Similar species Black Swift is larger, blackish
with a longer forked tail.
Voice Silent.
Habitat and behaviour Hawks insects during
migration flights along coasts, over open ground
and forest canopy, usually with swallows.
Range Breeds in eastern North America,
winters in western South America. Rare on
passage through the Bahamas, Cuba, Jamaica,
Hispaniola, Virgin Islands and the Cayman
Islands.
Status Irregular, rare to occasionally locally
abundant, passage migrant, August–November
and April–May.

Adult has large head, long wings and short tail with
spines.

Ruby-throated Hummingbird
Archilochus colubris

Taxonomy Monotypic.
Description L 9cm (3.4in).
Adult has iridescent green-bronze crown and upperparts, dark wings, white post-ocular spot, white breast and greenish-white underparts and long black bill; tail with white-tipped outer rectrices and black subterminal band projects slightly beyond wingtips; legs and feet dark. Male has brilliant iridescent red throat (looks black unless in sunlight) and notched tail. Female has black eye-line from lores to ear coverts and whitish speckled throat; immature similar to female.
Voice Rapid, high tics.
Habitat and behaviour Exotic flowering trees and urban/littoral gardens.
Range Breeds in North America; winters from Mexico to Panama when some cross

Adult male has iridescent green crown and upperparts and red throat, May.

the 1,000km Gulf of Mexico; rare on passage in the Bahamas, Cuba and Cayman Islands; vagrant elsewhere in the Greater Antilles. The Cayman Islands are one of only two island groups in the West Indies where hummingbirds do not breed.
Status Rare winter and passage migrant, December–April, observed on Grand Cayman only. First confirmed in 2005 despite persistent reports since the 1960s, the majority of which were misidentified as the moth *Aellopus tantalus*.

Adult female has white throat and white tips to outer tail feathers, February.

Belted Kingfisher
Megaceryle alcyon

Adult male has single grey breast-band, December.

Local name Kingfisherman.
Taxonomy Monotypic.
Description L 26–36 cm
(11–14 in). Adult has blue-grey
upperparts, large head with
shaggy crest; very large, long
heavy blackish bill, wide white
collar around throat and nape,
underparts white with blue-
grey breast-band, and short
tail. Male has blue-grey breast-
band only; female also has
rufous abdominal breast-band
and rufous flanks.
Similar species None.
Voice Harsh rattle *kek-kek-
kek-kek*.
Habitat and behaviour
Individuals hold winter
territories. Perches on trees and
telegraph wires along edge of
shores, mangrove lagoons and
herbaceous wetlands; hovers
and plunge-dives for fish, also
takes crabs, dragonflies and
one seen beating a land-crab
on the ground; flight erratic
with deep wing-beats.

Range Breeds in North
America; winters in south-west
United States, Middle America
to northern South America, and
the West Indies.
Status Fairly common winter
visitor and passage migrant,
August–early June.

Female shows grey upper- and rufous
lower breast-bands, March.

Adult male showing white patch
on outer primaries and white
underwing-coverts, March.

West Indian Woodpecker
Melanerpes superciliaris

Local Name Red-Head.
Taxonomy Polytypic (5).
Description L 26cm (10in).
Male is brilliant red from mid-crown to nape, and has red nasal tufts. Female has buffy crown and red nape. Both adults have finely barred black and creamy-buff upperparts and rump, uppertail-coverts white marked with black; pinkish-grey face, sides of neck and throat, rest of underparts buffy-cinnamon, red patch on lower abdomen (often not visible), black chevrons and bars on whitish-buff flanks and undertail-coverts. Long blackish bill with curved culmen, tail pointed and barred black and white, legs grey. Juvenile similar, dull barring on upperparts, underparts buffy-grey.
Similar species Yellow-bellied Sapsucker is smaller with black and white striped head and broad white patch on wing-coverts.

Adult male has red crown and nape, Grand Cayman only, March.

Adult male displaying gold and black back and coverts, black primaries, and black and white secondaries and central tail feathers, July.

Juvenile male in first flight, Grand Cayman only, July.

Adult female has red nape, Grand Cayman only, April.

Voice *Churr* and *krruuu-kruu-kruu*, continuous rapid *ke-ke-ke ke*, and call *key-ou*.

Habitat and behaviour Forages for arthropods, tree-frogs and many types of fruit by pecking and gleaning in all habitats from mid-levels to near the ground. Roosts in artificial bat houses. Nests, January–August, in living and dead tree cavities excavated by male and female, two broods regular, 2–5 eggs, 3–4 young usual, in dry and mangrove forest, often Black Mangrove trees and in Royal, Coconut and Silver Thatch Palms in urban/littoral areas. Often nests in the same cavity year after year. **Range** West Indian endemic with resident subspecies on Cuba, Isle of Youth, and Grand Cayman and two on the Bahamas (probably extirpated on Grand Bahama). **Status** *M. s. caymanensis*, an endemic subspecies, is a fairly common breeding resident on Grand Cayman only.

Adults feeding juvenile male, preparing to leave nest, July.

Yellow-bellied Sapsucker
Sphyrapicus varius

Taxonomy Monotypic.
Description L 20–23cm (8–9in). Male has a red chin and throat. Female has a white throat. Both sexes have red crown (restricted on female) bordered by black band, white nape, black and white facial stripes, black moustachial stripe joins black breast-band; back and wings heavily barred black and white, white wing-coverts form long wide patch on closed wing, underparts yellowish with blackish chevrons on sides and flanks. Short, straight black bill, black and white tail with black outer webs. Juvenile and first year are olive-brown, head and crown with buff streaks, upperparts black and buff.
Similar species West Indian Woodpecker is larger with red crown and nape, no white on wings.

Adult male has red throat, September.

Voice Silent.
Habitat and behaviour Forages for arthropods, ants, fruits and drills holes around tree boles, called sap wells, to drink sap (preferred trees include Red Birch, West Indian Almond); in littoral and urban areas in Coconut Palms, open woodland, and dry forest and Black Mangrove forest. Probably more common than records suggest as a considerable number of trees are used in successive years.
Range Breeds in northern North America; winters in North America south of the breeding range to Middle America and Panama and the West Indies in the Bahamas and Greater Antilles (rare east of Hispaniola), and Cayman Islands.
Status Uncommon winter visitor, August–April, with majority females and immatures.

Adult female has white throat, February.

Juvenile male developing red on throat, January.

Northern Flicker
Colaptes auratus

Adult male in breeding plumage with black malar stripe; underwing and undertail golden yellow, Grand Cayman only, April.

Local name Black Heart.
Taxonomy Polytypic (9).
Description L 30cm (12in).
Male has broad black
moustachial stripe and red
hind-crown and nape. Female
has grey crown with scarlet
triangle on central nape. Both
adults have cinnamon-buff
face, sides of neck, chin and
throat; long pointed black bill,
black upper breast crescent,
rest of underparts pale yellow
to pale cinnamon-buff with
bold black spots. Back and
wings brownish-grey barred
with black, rump and uppertail-
coverts white, heavily spotted
with black; tail black with shafts
of rectrices bright yellow and
outer rectrices barred black and
buff. Underwing-coverts and
undertail-coverts and undertail
show golden yellow in flight.
Juvenile is smaller and paler,
scarlet on nape replaced by
brownish barely visible short
feathers, moustachial stripe
orange in male.
Similar species None.
Voice *Kee-yar*, descending;
wicka-wicka repeated
quietly; long loud fast series
wikwikwikwik, also drumming
on tree bole.
Habitat and behaviour Forages

Adult female lacks black moustache, Grand Cayman only, October.

for arthropods (mainly ants), and wide variety of fruits, seeds and berries in all habitats from mid-levels down and often on the ground. Monogamous, breeds in tree cavities, often dead palms, January–August, in Mangrove and dry forest, single trees in pasture and edge of urban areas, and littoral areas (less frequently than the West Indian Woodpecker); 4–6 white eggs; young brooded and fed by both parents, fledge around 28 days.

Range Breeds in North America, Mexico, Cuba and the Cayman Islands. The *chrysocaulosus* group is endemic to the Greater Antilles and resident in Cuba, Isle of Youth and cays, and Grand Cayman.

Status Resident, endemic subspecies *C. a. gundlachi* is fairly common on Grand Cayman only.

Male showing black moustache and yellow shafts to feathers, April.

Caribbean Elaenia
Elaenia martinica

Adult showing white crown-patch on raised crest, May.

Local name Judas, Top-Knot Judas
Taxonomy Polytypic (7).
Description L 15–18cm (6–7in). Greyish-olive upperparts, darker on crown; white crown patch shows clearly when crest is raised, dull white orbital ring, two dull whitish wing-bars and whitish edging to coverts; chin, throat, breast and sides greyish-white, yellowish wash on lower abdomen and undertail-coverts, bill has pale lower mandible and often pale basal part to upper mandible. Juvenile is drab with brownish upperparts, buffy wing-bars, underparts grey-buff with yellow wash absent.
Similar species Acadian Flycatcher has pronounced white eye-ring. Eastern Wood Pewee has dark breast-band.
Voice Whistled *ph-eueer;* song *pee-wit-peeer* repeated or followed by *pee-weeew; peewit-peewit,* and a dawn trill.
Habitat and behaviour All terrestrial habitats; most common landbird in dry shrubland after the Bananaquit. Major part of diet is fruit and berries, followed by insects, and other arthropods. Aggressive behaviour and dominance in interspecific confrontations observed in breeding season. Shallow cup nest lined with spiders' webs and feathers sited 1.5–6m above the ground in all habitats (often in Bromeliads and Banana Orchids), 2 eggs buffy-orange with dark marking usual, female incubates and both parents feed young, 2–3 broods may be raised. Breeds February–October on Grand Cayman and

April or May–August on Little Cayman and Cayman Brac.

Range *E. martinica* is resident with disjunct distribution in the eastern and western Caribbean. Sedentary endemic subspecies occur on islands off the Caribbean coast of Belize and the Yucatan, Providencia, San Andreas, Netherlands Antilles, and Cayman Islands, and eastwards from Puerto Rico south through the Lesser Antilles.

Status Endemic subspecies *E. m. caymanensis* is a very common resident on Grand Cayman, east of Savannah; it is less conspicuous from late November–early January when it retreats into dense habitats. Very common on Cayman Brac and on Little Cayman where it is the only resident flycatcher.

Adult has pale lower mandible and two whitish wing-bars, April.

Two newly fledged juveniles with adult, July.

Eastern Wood-Pewee
Contopus virens

Taxonomy Monotypic.
Description L 14–16cm (5.5–6in). Adult has dark olive-brown upperparts, two broad whitish wing-bars on long olive wings, faint eye-crescent behind eye, orange lower mandible, dark grey breast almost forms a breast-band, whitish throat, yellow wash in centre of abdomen, and slightly forked tail.
Similar species Caribbean Elaenia has uniform whitish-grey underparts (yellowish wash on abdomen).
Voice Silent.
Habitat and behaviour Takes insects on the wing before returning to same perch; in urban areas, inland and littoral dry forest and shrubland, and Logwood on Grand Cayman.
Range Breeds in eastern North America; winters in South America. Migrates mainly through Middle America and casually (mainly in fall) through

Adult, note upright pose, long wings and tail, October.

the Bahamas, western Cuba and, rarely, Jamaica.
Status Uncommon to locally fairly common fall passage migrant, September–December, uncommon in spring, February–early May. One January record on Grand Cayman.

Adult showing pale lower mandible, grey breast-band and yellow wash on abdomen.

La Sagra's Flycatcher
Myiarchus sagrae

Adult breeding with primaries and uppertail-coverts edged dull rufous, Grand Cayman only, April.

Local name Tom Fool.
Taxonomy Polytypic (2).
Description L 19–22cm (7.5–8.5in). Ault has clove-brown head with slight crest, flattened forecrown, back brownish-olive, two white wing-bars with outer webs of primaries edged dull rufous, whitish on secondaries, throat and breast greyish-white, abdomen and undertail-coverts whitish, rump and uppertail-coverts grey, long tail with varying amounts of cinnamon on outer rectrices, large black bill, dark legs and feet. Breeding adult has lower back, rump and uppertail-coverts cinnamon-rufous and yellowish wash on lower underparts. Juvenile has rusty upperparts, wings and tail, and greyish underparts.
Similar species Great Crested Flycatcher perches vertically and abdomen is bright yellow. Gray and Loggerhead Kingbirds are larger and lack wing-bars.
Voice Song long rolling *brrrr-eeerr*; call *whuit*, alarm *weet-wee-weer,* and clicks bill.
Habitat and behaviour
Understorey and edges of dry and mangrove forest, in dry shrubland, usually away from habitation. Very inactive and approachable, perches with body slanted off vertical on mid

Adult showing bushy crest and black bill, April.

Adult showing dull rufous edges to rectrices, Grand Cayman only, April.

to low branches. Takes insects by sallies from a perch and by hover-gleaning for arthropods and larvae on leaves, fruit forms a large component of diet. Breeds, January–July; places cup nest in holes or cracks of trees, fence posts, old woodpecker holes; clutch 2–4 whitish eggs spotted with brown; both parents feed young.
Range Resident endemic subspecies, *M. s. sagrae*, in Cuba and the Isle of Youth, and Grand Cayman; *M. s. lucaysiensis* is resident in the Bahamas and absent from the Turks and Caicos Islands; casual in Florida.
Status Fairly common east of Savannah and uncommon to absent westwards, Grand Cayman only.

Eastern Kingbird
Tyrannus tyrannus

Taxonomy Monotypic.
Description L 23cm (9in). Head black, upperparts blackish-grey with two indistinct wing-bars, white underparts with grey wash on breast, square tail black with wide white terminal band, short black bill.
Similar species Loggerhead Kingbird has larger head, longer heavier bill, back is washed olive, yellow wash on lower abdomen, and tail has narrow white terminal band; Gray Kingbird has grey upperparts and dark facial mask.
Voice Silent.
Habitat and behaviour Littoral/urban areas, disturbed woodland.
Range Breeds in North America. Winters in South America; on passage mainly through Middle America, the northern Bahamas and uncommonly in western Cuba and Cayman Islands.
Status Short-stay passage migrant, rare to locally abundant in fall, September–October, and rare in spring in April; one January record on Cayman Brac.

Adult has smaller bill than other kingbirds and wide white tip to tail.

Gray Kingbird
Tyrannus dominicensis

Taxonomy Polytypic (2).
Description L 22–25cm
(8.5–10in). Grey head and
upperparts, orange crown
patch seldom visible, blackish-
grey mask from lores across eye
to ear-coverts, dark grey wing-
coverts edged pale, whitish
underparts with grey wash
on breast, heavy long black
bill, grey tail notched. Juvenile
washed brownish-grey on back
and wings.
Similar species Loggerhead
Kingbird has blackish head
without mask, and blackish
square tail with white tip.
Eastern Kingbird has dark
upperparts, square tail with
wide white terminal band.
Voice Song, rolling 5–6 syllable
pitch-ir-reir-rie-ee, shorter *pit-
tirrr-re*, call *peet*.
Habitat and behaviour
Dry shrubland, disturbed
woodland, littoral/urban
areas and rough pasture with
emergent trees; absent from
closed dry and mangrove
forest. Takes large flying insects
and lizards from an exposed
perch or telegraph wire; fruit

Adult has dark lores and ear-coverts, April.

forms a large part of diet,
especially Red Birch for large
flocks of arriving migrants.
Aggressive and territorial,
attacking frigatebirds and
raptors in flight. Non-breeding
birds form communal roosts
on Grand Cayman and a small
population breeds. Breeds,
April–July, nest is a large loose
cup nest lined with soft plant

fibre in trees and house gutters,
clutch 2–4 whitish or pink eggs
spotted with brown and grey.
Range *T. d. dominicensis* breeds
on south-eastern United States
Gulf coast, the Bahamas, Turks
and Caicos Islands, Greater
Antilles, Cayman Islands,
Trinidad, Netherlands Antilles,
and northern Venezuela. The
western populations of this
subspecies are migrant breeders
in the Bahamas, Cuba, Cayman
Islands and Jamaica wintering
in Panama and northern
South America; the eastern
population is resident from
Hispaniola east to the Virgin
Islands. *T. d. vorax* is resident in
the Lesser Antilles.
Status Summer breeding visitor,
mainly March–October but
occasionally February–December.
Fairly common, majority non-
breeding, on Grand Cayman
though occasionally fairly
common breeder; common and
breeding on Little Cayman and
Cayman Brac.

Adult; note long, heavy, broad black bill, April.

Loggerhead Kingbird
Tyrannus caudifasciatus

Adult, April.

Adult. Yellow edge to folded wing shows as yellow underwing-coverts in flight, October.

Local Name Tom Fighter.
Taxonomy Polytypic (7).
Description L 24–26cm
(9.5–10in). Adult has greyish-black head with yellow crown patch usually concealed; long, heavy, flattened, black bill with bristles at base, dark olive-brown back; brownish-grey wing-coverts edged whitish, may form indistinct wing-bars; yellow edge to folded wing and yellow wash on underwing-coverts; dark olive square tail with rufous uppertail-coverts and narrow whitish terminal band; throat and breast whitish, or white, with yellow wash on lower abdomen and undertail-coverts. Juvenile has greyish-brown upperparts, buff edges to wing-coverts,

and tail tipped orange-brown;
immature retains buffy wing-
coverts and tail tip.

Similar species Gray Kingbird
has pale grey head, black
mask and notched tail. Eastern
Kingbird has darker upperparts,
white underparts, smaller bill
and broad white terminal band
on square.

Voice Loud rolling call *treeerrrrr,*
and *pit-pit-pit-tirr-ee-eee.*

Habitat and behaviour
Mangrove and dry forest and
shrubland, urban and urban/
littoral areas, and single trees
in rough pasture on Grand
Cayman; mainly in dry forest
on Cayman Brac where it is
displaced from coastal urban
areas by Gray Kingbird in
summer breeding season.
Forages for large insects and

First-year, July. Note rufous on forecrown, and buff edges to wing-coverts and tip of tail.

lizards, hawking from a perch or
telegraph wire; also takes fruit
and berries. Breeds January–
September on Grand Cayman
and February–June on Cayman
Brac in a large, rough, unlined
cup nest, sited 3–10 m above
the ground in a tree fork on an
outer branch; 2–3 reddish eggs
marked with brown and violet;
young fed by both parents.

Range West Indian endemic
and Restricted Range species,
with resident subspecies in the
northern Bahamas, Greater
Antilles and Cayman Islands.
Casual in Florida.

Status Resident, endemic
subspecies *T. c. caymanensis*
is fairly common on Grand
Cayman and uncommon on
Cayman Brac. It has been
absent from Little Cayman since
the mid 20th century although
occasionally reported there.

Adult has yellowish undertail-coverts and square tail with white tip, February.

Scissor-tailed Flycatcher
Tyrannus forficatus

Adult male. Rare passage migrant, March.

Taxonomy Monotypic.
Description L 19–38cm (8–15in) including rectrices. Male has head and upperparts pale pearly-grey; very long. deeply forked, blackish tail with white sides; wings blackish edged whitish; salmon-pink wash on flanks, lower abdomen and undertail-coverts, axillaries reddish. Female duller, pink reduced and tail 30% shorter.

Juvenile/immature has buffy wash on abdomen and axillaries and shorter tail.
Similar species Fork-tailed Flycatcher has black head, dark grey back and entirely white underparts.
Voice Silent.
Habitat and behaviour Open dry shrubland, edge of urban areas with trees, perched on fences or telegraph wires.

Range Breeds in southern United States and northern Mexico, with small population in the Florida Cays; winters in Florida and Middle America to Panama; rare or vagrant in the West Indies in the Bahamas, Greater Antilles and Cayman Islands.
Status Rare passage migrant or vagrant, October–November.

Fork-tailed Flycatcher
Tyrannus savana

Taxonomy Polytypic (4).
Description L 33–41cm (13–16in). Large tyrannid with black head, flight feathers and tail; grey mantle and wing-coverts; white throat, partial nuchal collar and underparts. Adult male has forked tail with very long dark tail-streamers with narrow white edges, shorter in female. Juvenile has brownish wash on head and mantle and shorter tail.
Similar species Scissor-tailed Flycatcher has pale grey head and back; pink lower abdomen, underwing- coverts and undertail-coverts.
Voice Silent.

Habitat and behaviour. Catches insects on the wing in urban areas and over dry shrubland.
Range Breeds in Middle and South America; nomadic and partly migratory; some winter in South America and irregularly to the southern Lesser Antilles; casual in western Cuba, Jamaica and northern Lesser Antilles.
Status Vagrant or rare fall passage migrant, October–November; one on Little Cayman, January 2006.

Adult male. Note white collar and white sides to tail, November.

White-eyed Vireo
Vireo griseus

Taxonomy Polytypic (7).
Description L 12.5cm (5in).
Adult has grey head and nape,
greenish upperparts and rump
with yellowish wash, bright
yellow eye-ring and lores
forming 'spectacles', white
iris is diagnostic, dark wings
with two pronounced whitish
or yellowish-white wing-bars,
whitish throat and underparts
with sides washed yellow.
Immature and juvenile have
dark iris and spectacles are
buffy.
Similar species Thick-billed
Vireo has dark iris and yellowish
underparts. Yellow-throated
Vireo has dark iris; yellow chin,
throat and breast, and white
abdomen.
Voice Song similar to Thick-
billed Vireo, 3–7 syllables
beginning and ending in *chic*
but higher pitched, faster and
sharper, calls *churr-churr* and
pick.
Habitat and behaviour Usually

Adult has yellow 'spectacles', white iris and whitish throat, February.

concealed in fairly dense
understorey of woodland,
dry shrubland and mangrove

shrubland, foraging slowly
among leaves for arthropods
(mostly insects), seeds and
fruits including Red Birch,
Pepper Cinnamon, Wild
Fig; curious and briefly
approachable.
Range Breeds in eastern
United States and Mexico;
V. g. noveboracensis winters
in Middle America and the
Bahamas and western Greater
Antilles in Cuba (where *V. o.
griseus* is also a rare migrant)
and Cayman Islands; vagrant or
rare on passage elsewhere.
Status Uncommon passage
migrant, September–December
and February–April; one
January record.

Adult has grey head, two wing-bars
and greenish back and rump,
February.

Thick-billed Vireo
Vireo crassirostris

Taxonomy Polytypic (5).
Description L 13.5cm (5.25in).
Adult has head greyish olive;
back, rump and uppertail-
coverts olive-green; wings
and tail greyish-brown with
pale olive edgings; middle
and greater wing-coverts
broadly tipped white forming
two wide white wing-bars;
broken yellowish eye-ring and
yellow above lores forming
'spectacles', greyish-black lores,
iris light brown. Underparts
including undertail-coverts vary
from pale yellowish-buff to
pale yellow, becoming brighter
yellow in the breeding season,
underwing-coverts pale yellow;
bill heavy, flattened, dark
and slightly hooked; legs and
feet blue-grey. Juvenile and
immature have brownish wash

Adult; note dark lore, Grand Cayman and Cayman Brac only, February.

Adult has heavy flattened bill and pale edgings to wings, April.

on upperparts, lack blackish
lores; underparts buffy-olive
and wing edgings dull buff.
Similar species White-eyed
Vireo has white iris, smaller
bill, yellow sides and rest of
underparts whitish. Yellow-
throated Vireo has yellow
chin, throat and breast and
contrasting white abdomen.
Voice Song, a long, assertive,
fast, 5–9 varied syllables; often
beginning and/or ending in
chic, similar to White-eyed
Vireo but slower, lower pitched
and less harsh; calls, sharp *chic*
and alarmed growling *churr-
churr-churr.*
Habitat and behaviour
Forages for arthropods slowly
examining branches and leaves,
also takes fruit, in open areas
of dry forest understorey, dense
shrubland, and Logwood on

Adult developing breeding plumage when underparts and wing edgings become brighter yellow, February.

Adult showing yellow 'spectacles', February.

Grand Cayman; absent from mangrove and Buttonwood. Usually paired and very aggressive with other vireos. Males begin to sing from late February and breed April–July. Deep pendant cup nest is suspended from a branch by strips of vegetation, lined with leaves, bromeliad down and strips of red birch, low down in dry shrubland; monogamous, female incubates 2 whitish-buff eggs with heavy dark spots and both parents feed young.
Range Restricted Range species. West Indian endemic with resident subspecies on the Bahamas, Caicos Islands and cays off north-east Cuba (mainly Paredon Grande), Cayman Islands, Tortue Island off Hispaniola, Providencia and Santa Catalina islands.
Status Endemic subspecies *V. c. alleni* is resident on Grand Cayman and Cayman Brac, and extirpated on Little Cayman (although reported several times and may become re-established). Fairly common in the eastern half of Grand Cayman, becoming locally common in the north and central districts, but due to its sensitivity to urbanisation immediately shows declines in areas of new urban development. It is common on Cayman Brac, becoming locally very common on the central bluff.

Active pendant nest attached on one side to branch, May.

Yellow-throated Vireo
Vireo flavifrons

Taxonomy Monotypic.
Description L 12.5cm (5.5in).
Adult has crown olive-green,
back greenish becoming grey
on lower back and rump, two
sharply defined white wing-
bars, yellow 'spectacles' around
eye and above lores, lores
fainter than Thick-billed Vireo,
iris dark; chin, throat and breast
yellow, abdomen and undertail-
coverts whitish. Juvenile has
upperparts brownish, with
buffy-yellow throat and breast.
Similar species Thick-billed
Vireo has pronounced black
lores, no white on underparts.
White-eyed Vireo has white iris
and whitish throat and breast.
Voice Harsh alarm call heard
infrequently.
Habitat and behaviour Inland

Adult. Only vireo with yellow throat and breast and white abdomen, December.

woodland and dry shrubland.
Range Breeds in eastern North
America; winters from Middle
America to northern South
America and the West Indies,
where it is regular in the
Bahamas and Greater Antilles,
except Jamaica and Puerto Rico
where it is rare.
Status Uncommon to locally
fairly common winter visitor,
September–April.

Red-eyed Vireo
Vireo olivaceus

Taxonomy Polytypic (12).
Description L 15cm (5.75in).
Adult has grey crown bordered
by two lateral dark stripes,
long whitish supercilium, dark
eye-line, dark red iris. Wing-bars
absent. Upperparts olive-grey,
underparts whitish with greenish
wash on sides, flanks and
undertail-coverts. Juvenile has
brown iris in fall, yellow wash on
flanks and undertail-coverts.
Similar species Black-whiskered
Vireo has black malar stripe.
Yucatan Vireo and Philadelphia
Vireo have no lateral crown
stripes and the latter has yellow
chin, throat and breast.
Voice Silent.
Habitat and behaviour
Migrants forage at all levels
from shrubs in coastal areas,

Adult. Note black stripes on grey crown, long eye-stripe and long supercilium.

canopy of mangrove and dry
forest to gardens in urban areas;
take arthropods and fruits.
Range Breeds throughout the
americas; *V. o. olivaceus* winters
in South America and occurs on
passage in the West Indies in the
Bahamas, western Greater Antilles,
and Cayman Islands; vagrant
elsewhere. Eight subspecies
resident in South America.
Status Very uncommon to
locally fairly common (when
associated with tropical storms)
passage migrant, August–
November, rare in spring.

Black-whiskered Vireo
Vireo altiloquus

Taxonomy Polytypic (6).
Description L 15–16.5cm
(5.75–6.5in). Large vireo with
long pointed bill and long
wings, black malar stripe
('whisker') is diagnostic but
may be obscured. Wing-bars
absent. Adult has dark grey
crown bordered by blackish
lateral stripes, long greyish-
white supercilium, eye-line
greyish, iris red, upperparts
brownish-olive, wings and tail
edged yellowish in breeding
season, throat and underparts
ivory with yellowish wash on
sides, flanks and undertail-

Adult showing black 'whisker', Little Cayman and Cayman Brac only, April.

coverts. Juvenile has brownish
upperparts, dark iris and faint
wing-bars and malar stripe.
Similar species Red-eyed,
Philadelphia and Yucatan Vireos
lack malar stripe. Yucatan and
Philadelphia Vireos also lack
crown-stripe.
Voice Song loud, clear,
repeated single and double
couplets *good-john, two-whit*,
call *chew-we* and *chew-chew*.
Habitat and behaviour Forages
for arthropods, seeds and fruits.
On Little Cayman and Cayman
Brac migrants arrive in April,

breed May–July. Deep cup nest
suspended between a branch
fork, pendant, anchored by
strips of fibre and birch bark,
lined with leaves, spider's
webs, in dry forest canopy;
monogamous, female broods
2–3 whitish eggs spotted at
broad end, and both parents
feed young. Migrants on Grand
Cayman occur in dry and
mangrove forest and shrubland,
and urban/littoral areas.
Range *V. altiloquus* breeds in
Florida, Netherlands Antilles,

Trinidad and islands off
Venezuela and the West Indies
including Swan Island and
Providencia; winters in northern
South America from Colombia
to Brazil. *V. a. barbatulus* is
a summer breeding visitor in
southern Florida, the Bahamas,
Cuba, Little Cayman and Cayman
Brac; on passage only on Grand
Cayman. *V. a. altiloquus* is a
summer breeding visitor on
Jamaica, Hispaniola (where some
are resident) and Puerto Rico;
resident in the Lesser Antilles.
Status *V. a. barbatulus* is an
uncommon summer breeding
visitor on Little Cayman and
locally fairly common on the
bluff on Cayman Brac, January–
September. Uncommon to fairly
common on passage on Grand
Cayman, September–October
and, more frequently, January–
May; possible breeding has not
been ruled out as a few singing
males have been heard. *V. a.
altiloquus*, collected on Grand
Cayman in April 1974, was
probably an overshoot from
Jamaica and this subspecies
may also occur regularly.

Breeding adult showing wings and tail edged with olive-yellow, April.

Yucatan Vireo
Vireo magister

Adult has slate eye-line, whitish underparts with yellowish–olive wash on sides and flanks, Grand Cayman only, April.

Local name Sweet Bridget.
Taxonomy Polytypic (4).
Description L 15cm (5.75in).
Adult has crown and hindneck greyish-olive, becoming greenish-olive on back, rump, uppertail-coverts and scapulars; pale yellow underwing-coverts, undertail-coverts and axillaries; wing and tail feathers edged bright olive, inner webs of tail and wing feathers brown. Wing-bars absent. Dull white supercilium broadest near eye, iris brown, lores and post-ocular line slate grey, paler behind eye; ear-coverts greyish-buff, sides of face buffy-white becoming grey on sides of neck. Malar area, chin, throat and breast whitish, pronounced yellowish-olive wash (becomes yellow for a short period early in the breeding season) on abdomen, sides and flanks. Stout, heavy bill greyish with lower mandible paler, legs greyish-blue. Juvenile has browner upperparts, buffy

supercilium and underparts.
Similar species Red-eyed and Black-whiskered Vireos have red iris and grey crown with black lateral stripes. Black-whiskered Vireo has black moustache

stripe. Philadelphia Vireo (vagrant) has brown iris, yellow throat and breast.
Voice Song 3-syllable *sweet-bridg-et* and 2-syllable whistled *whoi-whu*.

Adult, April.

Habitat and behaviour Forages in all habitats, most commonly in the canopy of dry forest and black mangrove forest in the Mastic Reserve and the edge of the Central Mangrove Wetland. Food includes arthropods and vegetable matter. Territorial males sing from February to November, courtship begins in January and breeding (when many are nests built and abandoned) March–September. Nest similar to Thick-billed Vireo, from 3m but usually close to or in the canopy; a deep pendant cup of palm fibre, birch bark and spiders' webs lined with fibre and suspended from a branch fork, anchored by strips of palm fibre. Monogamous, female incubates two whitish eggs spotted red at the large end and both parents feed young.
Range Restricted Range species. Three subspecies breed in coastal Yucatan, Honduras and Belize including the offshore islands and cays, and a fourth on Grand Cayman.

Juvenile has paler plumage and rectrices are tapered, May.

Status Endemic subspecies *V. m. caymanensis* is a common to locally very common resident on Grand Cayman, east of Savannah; rare west of Savannah due to habitat loss.

Adult sitting in pendant nest, May.

Purple Martin
Progne subis

Adult male is glossy blue-black overall.

Taxonomy Polytypic (3).
Description L 19cm (8in). Male is glossy metallic blue-black overall. Female duller purple-blue, forecrown and hindneck grey; throat, upper breast and sides scaly with brown streaks edged with grey, rest of underparts whitish-grey. Tail moderately forked. Juvenile has grey-brown upperparts and breast, and greyish-white underparts.
Similar species Adult Caribbean Martin has white lower breast and abdomen.
Voice Series of buzzy calls and chirps.
Habitat and behaviour Over the coastal areas and open ground; large flocks occasionally perch on overhead wires.
Range Breeds in North America and Mexico; winters in South America; majority migrate through Middle America, and *P. s. subis* also occurs the West Indies: in the Bahamas (uncommon), Cuba (common) and Cayman Islands.
Status Uncommon to occasionally locally abundant on passage mainly in fall, July–October, and March–May.

Adult female, April.

Caribbean Martin
Progne dominicensis

Adult male has white lower breast, abdomen and undertail-coverts.

Taxonomy Monotypic.
Description L 20cm (8in). Male has black wings and tail, glossy steel-blue upperparts, throat, upper breast and sides; white lower breast, abdomen and undertail-coverts. Female and juveniles have upperparts brown, wings and tail grey-black with blue wash; throat pale grey-brown, darker on breast, sides and flanks. Tail moderately forked.
Similar species Purple Martin male is entirely blue-black; female has grey hind-collar, brownish-grey scaling on throat and upper breast.
Voice Silent.
Habitat and behaviour Seen singly with Purple Martin or swallows over coastal areas.
Range Breeds in the West Indies from Jamaica and Hispaniola eastwards through the Lesser Antilles; winter range unknown, probably South America; casual or vagrant in the Bahamas and Turks and Caicos Islands.
Status Rare passage migrant, February–early June and December.

Adult female, June.

Tree Swallow
Tachycineta bicolor

Adult males. Only swallow with blue-green upperparts combined with white underparts, February.

Taxonomy Monotypic.
Description L 14cm (5.5in).
Adult has glossy blue-green
upperparts, black mask from
bill to side of neck, blackish
wings and tail, and white
underparts; in flight shows
white extending on to lower
back. Male brighter, female
may be duller with grey wash
on breast. Juvenile and first
year have brown upperparts
and grey-brown wash on
underparts. Tail notched.
Similar species Only swallow
with blue-green upperparts.
Voice Silent.
Habitat and behaviour
Seen on passage with other
swallows, hawking over open
ground and the shore.
Range Breeds in North
America; winters from southern
North America through Middle
America to northern South
America and the West Indies:
in the northern Bahamas,
Cuba, Hispaniola, Jamaica and
Cayman Islands.
Status Regular, uncommon
to locally abundant, short-
stay spring passage migrant,
January–May.

Adult male. May.

Adult female, February. Occasionally abundant in spring.

Northern Rough-winged Swallow
Stelgidopteryx serripennis

Adult has warm brown upperparts and head, December,

Taxonomy Polytypic (6).
Description L 12.5–14cm (5–5.5in). Adult is entirely warm brown above with paler rump, darker wings and tail; brownish throat, breast, sides and flanks, rest of underparts whitish. Tail slightly forked. Juvenile has pale rufous wash on upperparts, throat and upper breast.
Similar species Bank Swallow is grey-brown above with broad breast-band and white underparts.
Voice Call a series of soft *brrrt* repeated.
Habitat and behaviour Hawks for flying insects over open ground and wetlands, usually with Barn Swallows.
Range Breeds in North America and Middle America to Panama; winters from the United States to Panama; rare on passage in the Bahamas, Turks and Caicos Islands, and Greater Antilles in Cuba, Jamaica and Hispaniola, mainly in fall; very rare in the Virgin Islands.
Status Rare to occasionally fairly common passage migrant, July–May, mainly late August–November and March–April; records of single birds, December–February, and May.

Adult showing pale brownish throat, breast and sides, October.

Bank Swallow
Riparia riparia

Taxonomy Polytypic (5). Also known as Sand Martin.

Description L 12–14cm (5–5.5in). Small swallow with grey-brown upperparts, white underparts except for brown breast-band. Juvenile has pale feather edges and buffy wash on underparts. Tail has shallow fork.

Similar species No other swallow has a distinct breast-band and white underparts.

Voice Sharp twittering calls.

Habitat and behaviour Usually with Barn Swallows over open ground and the shore.

Range Breeds northern hemisphere. North American population of *R. r. riparia* winters in South America; rare on passage in the West Indies except in the Bahamas, Puerto Rico, Virgin Islands and Cayman Islands.

Status Uncommon to occasionally locally common passage migrant, September–early November, and March–May. Records of single birds with other swallows in all months except June.

Adult. Only swallow with brown breast-band and white underparts, September.

Cliff Swallow
Petrochelidon pyrrhonota

Taxonomy Polytypic (4).

Description L 13–15cm (5.25–5.75in). Adult has glossy dark blue crown with white forehead, pale grey collar, white streaking on dark blue mantle, blackish-brown wings and tail, buffy-cinnamon rump, dark rufous face, black or dark rufous chin and throat, buffy breast and sides, rest of underparts buffy-white. Square tail. Juvenile has dark forehead, face and throat.

Similar species Cave Swallow has rufous forehead and paler throat.

Voice Silent.

Habitat and behaviour Coastal and wetland areas with other swallows.

Range Breeds in North America and Mexico; winters in southern South America; on passage in the West Indies where rare in the Bahamas, Cuba, Virgin Islands and Cayman Islands; vagrant to Jamaica and Hispaniola or, more likely, overlooked among breeding Cave Swallows.

Status Rare on passage, October–November and February–May.

Adult showing white forehead, rufous face and blackish throat.

Cave Swallow
Petrochelidon fulva

Taxonomy Polytypic (6).
Description L 12cm (5cm).
Adult has dark blue crown;
rufous forehead, wide collar
and rump; white streaking on
dark blue mantle, rufous cheeks
and throat, buffy breast and
whitish abdomen. Tail almost
square.
Similar species Cliff Swallow
has white forehead and darker
throat.
Voice Silent.
Habitat and behaviour
Wetlands, coasts and urban
areas with Barn Swallows,
usually seen on the wing.
Range Populations of this
species complex breed in
southern-central United States,
Mexico, Ecuador, Peru and
Greater Antilles. Populations are

Adult at nest, Dominican Republic, May.

migratory on Cuba (breeds in
summer) and mostly sedentary
on Jamaica, Hispaniola and
Puerto Rico. Vagrant to the

Lesser Antilles.
Status Rare and irregular
passage migrant, September–
October and February–May.

Barn Swallow
Hirundo rustica

Adult male has dark rufous throat and breast and partial breast-band, May.

Adult male Barn Swallow has longer tail than female, April.

Taxonomy Polytypic (8).
Description L 15–19cm
(5.75–7.5in). Adult has rufous
forehead; throat, crown and
upperparts glossy steel-blue,
more iridescent in male,
blackish wings, tail deeply
forked with elongated outer
tail feathers and white tail
spots. Male has underparts
pale to rich cinnamon-rufous
and longer tail; female has
less glossy upperparts, paler
underparts and shorter tail.
Juvenile has brownish wash on
blue upperparts, pale cinnamon
throat, rest of underparts
whitish, and short tail.
Similar species No other adult
swallow has long tail streamers
and tail spots.
Voice Twittering song, and *wit-
wit* call while hawking.
Habitat and behaviour Usually
in flight over urban and open
areas, lagoons and coasts or
at rest on wires and on shores.
Flocks of exhausted and dehy-
drated birds fall out after storms
and cover roads and pastures.
Range Breeds in northern
hemisphere. *H. r. erythrogaster*
breeds in North America,
Mexico and Argentina; winters
in Middle and South America,
Puerto Rico and the Lesser
Antilles; on passage throughout
the West Indies.

Status Observed in all months.
Common to abundant on
passage, mid-March–early May
and September–November (in
exceptional numbers following
storms); uncommon in other
months.

Adult female has paler underparts than male, April.

Blue-grey Gnatcatcher
Polioptila caerulea

Taxonomy Polytypic (9).
Description L 11cm (4.25in).
Sexes similar in winter. Non-breeding adult is very small with conspicuous white eye-ring, bluish-grey upperparts, dark wings, whitish underparts, long fine bill and long black tail with white outer tail feathers (shows as white undertail when perched). Breeding male has black line from forehead curving over eye to ear-coverts, absent in female. Immature has brownish wash on upperparts.
Similar species None.
Voice Mewing call *zee* or *zee-zee*, softer than Gray Catbird.
Habitat and behaviour
Understorey of mangrove and dry forest and shrubland and Buttonwood edge; frequently holds tail vertically and fans tail.
Range Breeds in the United States to Mexico, the Bahamas and Turks and Caicos Islands;

Male in breeding plumage with black supercilium, February.

winters to Middle America, Cuba (common) and the Cayman Islands.

Status Increasingly rare passage migrant and winter visitor, August–April.

Adult non-breeding, October.

Veery
Catharus fuscescens

Taxonomy Polytypic (4).
Description L 16–18cm (6.25–7in). Medium-sized thrush. Adult upperparts vary from bright tawny to reddish-brown, pale indistinct eye-ring, white throat, indistinct malar stripe, fine brown spots on breast washed warm buff, flanks greyish and rest of underparts whitish; bill dark with pale lower mandible, legs flesh, short tail and long rounded wings. First year similar but pale edges to wing-coverts.
Similar species Less spotting on breast than other migrant thrushes, and upperparts brown not grey.
Voice Silent.
Habitat and behaviour Dry forest edge, woodland; Migrants are predominantly terrestrial, hopping in leaf litter (especially when damp) foraging for insects, including

Adult. Note reddish-brown upperparts and fine spots on buffy breast, May.

ants and termites, and gastropods.
Range Breeds in North America and winters in Brazil. On passage through Middle America and the West Indies in the Bahamas, Cuba, Jamaica and Cayman Islands, vagrant to Hispaniola.
Status Rare to locally fairly common passage migrant, September–October, and April–May; in fall migration occurs in mixed flocks with Swainson's and Gray-cheeked Thrushes.

Gray-cheeked Thrush
Catharus minimus

Taxonomy Polytypic (2).
Description L 16–20cm (6.25–8in). Grey-brown upperparts, grey cheeks, indistinct pale eye-crescent behind eye, dark malar stripe, whitish throat, heavy blackish-brown triangular spots on entire buffy breast becoming blurred on whitish abdomen, flanks olive-grey, and legs flesh. First year similar but wing-coverts edged pale.
Similar species Swainson's Thrush is larger, distinct buffy eye-ring and bar forming 'spectacles', spots on buffy breast are brownish.

Adult. Note grey cheeks and triangular spots on breast, April.

Voice Silent.

Habitat and behaviour Dry forest edge, open woodland, dense secondary growth woodland, gardens with dense bushes, predominantly terrestrial on migration, observed hopping in leaf litter foraging for insects; also takes fruit.

Range Breeds in northern North America and winters in northern South America. The subspecies *C. m. aliciae* occurs

Adult. Triangular breast spots become blurred on sides and abdomen, April.

on passage through Middle America, rarely in the western Greater Antilles (Cuba and Jamaica) and Cayman Islands. Replaced by Bicknell's Thrush *Catharus bicknelli* in Hispaniola. Both species occur in Cuba, Jamaica and Puerto Rico (but have not yet observed in the Cayman Islands).

Status Regular but rare passage migrant, October–November and April–May; often arrives in fall migration with mixed flocks of Swainson's Thrushes and Veerys.

Swainson's Thrush
Catharus ustulatus

Taxonomy Polytypic (6).

Description 17.5cm (7in). Olive-brown upperparts, buffy cheeks and throat, distinct creamy-buff eye-ring and bar above lores forming 'spectacles', dark malar stripe connects with brown triangular spots on breast becoming blurred on reddish-brown washed sides and flanks, bill dark with pale base to lower mandible, short legs flesh. First-year similar but wing-coverts edged buff.

Similar species Gray-cheeked Thrush has greyish upperparts and grey cheeks; that species and Veery lack pronounced eye-ring.

Voice Silent.

Habitat and behaviour Arrives in mixed flocks with other thrushes in fall and spreads throughout inland urban habitats, secondary woodland and open dry forest on Grand Cayman; high numbers often maintained for 4–5 weeks. Forages on the ground for

Adult has buffy eye-ring and line above lores forming 'spectacles'.

insects and insect larvae, especially ants, termites and grasshoppers, and takes fruits (Smilax, Pepper Cinnamon).

Range Breeds in North America; winters in Middle America and northern Argentina, *C. u. swainsoni* occurs on passage in the

Bahamas, Cuba, Jamaica and Cayman Islands.

Status Uncommon passage migrant, irregularly locally abundant in fall, mid September–early November, in mixed flocks with Veery and Gray-cheeked Thrush; also observed, late March–May.

Wood Thrush
Hylocichla mustelina

Taxonomy Monotypic.
Description L 20cm (8in). Large thrush with rufous-brown crown, nape and neck; darker rufous-brown on back and tail, pale lores and pronounced eye-ring, streaked ear-coverts, white throat and underparts, bold blackish spots on breast and flanks, legs pale flesh. Immature has brownish upperparts streaked with buff and spotted underparts.
Habitat and behaviour Dry forest edge with dense understorey, second growth woodland; terrestrial on migration hopping in leaf litter (especially when damp) foraging for fruit and insects, including ants, termites and small beetles.

Adult has white eye-ring and large black spots on breast and flanks, February.

Range Breeds in North America and winters in Middle America. Rare on passage and occasionally overwinters in the Greater Antilles.
Status Rare passage migrant, October–November and February.

Red-legged Thrush
Turdus plumbeus

Adult. Breeds on Cayman Brac only, February.

Adult. Note white chin, black throat and dark orange on lower abdomen, February.

Local name Old Truss.
Taxonomy Polytypic (6).
Description L 25–28cm
(10–11in). Large bluish
slate-grey thrush; sexes alike
though female may be smaller
and duller. Adult dark slate
on upperparts, wings black
edged with grey, blackish-grey
lores, coral red eye-ring, white
chin immaculate or crossed
by fine black lines extending
to white malar stripe, black
throat, bill red on upper
mandible, legs red. Lower
abdomen to vent has various
amounts of blackish orange
to rich cinnamon-orange,
undertail-coverts white, long
black tail with white tail spots
showing as white tips to outer
tail feathers in flight. Juvenile
duller, wing-coverts mottled
buffy and faintly streaked

blackish, underparts greyish, throat and upper breast spotted blackish, bill pale bone or yellowish.

Voice Complex; loud and long versatile song, series of 1-, 2- and 3-syllable fluting, grating, whistling phrases, with short pause between each note, and interspersed with insect-like *tsrrip;* one song in February lasted for 30 minutes. Calls *churr* and *pita* repeated, aggressive rapid *chee-chee-chee*, and very loud staccato alarm *wit-wit-wit;* a mimic, copying parrot, mockingbird and mobile phone calls.

Habitat and behaviour
Forages and nests in all habitats other than wetlands: from dry forest to urban areas and second growth. Forages from the canopy to leaf-littered karstic rock floor; omnivorous, taking arthropods (cockroaches, spiders, moths, larvae, ants), lizards and snakes; also many fruits including Lantana, Wild Fig, Pepper Cinnamon. Arboreal and highly visible in the breeding season, but thereafter often retreats to forest understorey, November to January. Perches to rest in trees for 2–3 hours in the middle of the day; holds tail overhead when landing, also droops wings. Very prone to foot pox (*Poxvirus avium*) leading to leg loss. Breeds mainly March–early July, occasionally to September. Highly territorial but constantly persecuted and dominated by Northern Mockingbirds in urban areas. Pairs are monogamous for at least a season; female builds nest, male defends, and both incubate. Bulky, loose cup nest lined with dried vegetation, birch bark and down of epiphytes, usually 3–8m up

Adult has black wings edged grey and long tail tipped white on outer rectrices.

in dry forest, often concealed in vines, and in Silver Thatch Palms in dry shrubland. Also in isolated trees in pasture on the bluff, littoral areas, and gardens and gutters of houses (at the same house year after year) on the coastal plain and the bluff. Clutch 2–4, pale greenish-blue eggs streaked reddish brown, hatch after *c.*12 days and both adults feed young. Nest predation by Smooth-billed Ani

and snakes frequent.

Range West Indian endemic with subspecies in the northern Bahamas, Cuba, Cayman Islands, Hispaniola, Puerto Rico and Dominica in the Lesser Antilles.

Status Endemic subspecies *T. p. coryi* is a fairly common breeding resident on Cayman Brac only. Occasional rare visitor to eastern Little Cayman to forage.

Gray Catbird
Dumetella carolinensis

Adult has black cap and brownish undertail-coverts, May.

Taxonomy Monotypic.
Description L 23cm (9in). Adult dark slate-grey above, paler below; forehead, crown and tail black, wings short, undertail-coverts orange-brown, and bill and legs slate grey.
Similar species None.
Voice Cat-like mewing *mea*, and short low-pitched *churr*.

Habitat and behaviour Usually concealed in understorey of all habitats, with dry shrubland preferred, and urban/littoral areas. Solitary and shy, heard more often than seen, except on migration when small flocks are usual. Flicks tail or holds it erect with wings drooped.
Range Breeds in North America. Winters to southern United States, the Caribbean slope of Middle America to northern South America and the West Indies, where it is regular in the Bahamas, Cuba, Jamaica and Cayman Islands and rare elsewhere in the Greater Antilles.
Status Fairly common winter visitor and common passage migrant, October–May.

Adult in characteristic pose with tail fanned and wings drooped, April.

Northern Mockingbird
Mimus polyglottos

Adult. Characteristic pose with tail raised and wings drooped, January.

Taxonomy Polytypic (3).
Description. L 23–28cm (9–11in). Crown, face and upperparts pale grey, underparts whitish-grey, dark grey eye-line, yellow or orange iris, broad dark grey wings, two white wing-bars, white patch on edge of closed wing shows as white base to primaries in flight or when wing flashing; long, blackish-grey tail, white underneath, shows as white outer tail feathers in flight; bill and legs blackish. Juvenile has pale greyish-brown back with greyish-brown spots on greyish breast, wing coverts edged buffy-grey, greyish iris.
Similar species None.
Voice Loud beautiful song, sustained series of short highly variable phrases, often sings at night, and mimics other species, telephones and ambulance sirens. Many calls, including a sharp *chaak*.

Habitat and behaviour Open disturbed habitats, urban and littoral areas, dry shrubland including xeric vegetation on exposed eastern bluffs, and edge habitats, but absent from interior of dry and mangrove forest. Omnivorous, feeding on fruits (Red Birch, Pepper Cinnamon, Wild Fig, and Papaya and Mango opened by parrots), and insects and lizards pursued over low vegetation or on the ground. Holds tail

Adult has yellow or orange iris, November.

Juvenile has spots on breast and grey iris, April.

erect and wings drooped or raises wings exposing white patches when wing- flashing. Highly territorial, holding winter territory. Monogamous by season, breeding December–September on Grand Cayman and to early August on Little Cayman and Cayman Brac. Rough open cup nest of twigs, built by male, lined with grasses at variable heights usually <3 m, aggressively defended with attacks on pets, people (inflicting stab wounds), birds: especially grackles and anis, and snakes; usually 2–3 blue-green eggs with dark brown marks, incubated by female; young fed by both parents and fledge at *c.* 12 days. Up to three broods are raised in some years.

Range Breeds in North America, Mexico and West Indies; *M. p. orpheus* breeds in the Bahamas, Greater Antilles and Cayman Islands.

Status Common on Grand Cayman, where one of the most conspicuous breeding residents. Also common on Cayman Brac and fairly common and increasing on Little Cayman. The population continues to increase in line with urban development.

Adult 'wing-flashing' showing white base to primaries, July.

Cedar Waxwing
Bombycilla cedrorum

Adult. Note red shaft extensions and yellow-tipped tail, February.

Taxonomy Polytypic (2).
Description L 18cm (7in).
Head with conspicuous crest;
cinnamon-brown crown,
back and breast; black mask
from base of bill across eyes
to nape outlined with white,
brown iris; rump and uppertail
pale grey, long wings have
grey-brown coverts with
dark flight feathers and red
shaft extensions at tips of
secondaries, pale yellowish
abdomen, undertail-coverts
white, short tail with yellow
terminal band, bill blackish,
legs dark brown. Juvenile has
streaked buffy underparts and
lacks shaft extensions on wing.
Similar species None.
Voice Hissing *szeeee* among
flock on take-off.
Habitat and behaviour
Irruptive migrant, flocks
in woodland and urban
littoral areas coinciding with
abundance of fruiting trees,
e.g., Pepper Cinnamon, Wild
Fig; also take insects.
Range Breeds in North
America; winters in North
America south through Middle
America to Panama, and
irregularly in the Bahamas,
Cuba and Cayman Islands;
casual in the rest of the
Greater and Lesser Antilles.
Status Uncommon to locally
abundant, irregular winter
visitor and passage migrant,
mainly on Grand Cayman,
January–May.

Ovenbird
Seiurus aurocapilla

Adult in leaf-litter with tail raised showing white undertail-coverts, April.

Taxonomy Polytypic (3).
Description L 14–16.5cm (5.5–6.5in). Large thrush-like terrestrial warbler with no wing-bars or tail spots. Adult has face and upperparts olive-brown to pale brown, orange central crown stripe bordered by black lateral stripes, dark iris with wide white eye-ring, black malar stripe, underparts white with blackish spots and streaks on breast to flanks, bill grey with pale lower mandible, and pinkish legs. First-winter similar with rufous wash on tail and tertials.
Similar species All other terrestrial warblers have a supercilium.
Voice Sings occasionally in late spring *teach-er teach-er*; call *chuup*.
Habitat and behaviour Bobs head and flicks tail, also carries tail erect with wings drooped while foraging in damp leaf-littered ground in all forest habitats and dry shrubland, and in gardens on passage. Small flocks occur on passage; otherwise solitary, holding winter territories.
Range Breeds in North America; winters in Florida, the Gulf coast, both coasts of Middle America to Panama, (casually to northern South America) and the West Indies: in the Bahamas, Greater Antilles, Cayman Islands, and rarely in the Lesser Antilles. The subspecies in the Cayman Islands include *S. a. aurocapilla* (majority); *S. a. furvior* and *S. a.cinereus* may also occur.
Status Fairly common winter visitor and passage migrant, August– May.

Adult showing median orange crown-stripe bordered by lateral stripes, April.

Worm-eating Warbler
Helmitheros vermivorum

Taxonomy Monotypic.
Description L 14cm (5.5in).
Adult stocky with flat crown
and short tail. No wing-bars or
tail spots. Olive-brown back,
cinnamon crown with black
lateral stripes, black eye-line to
nape, cinnamon-buff throat
and breast, undertail-coverts
buffy-white, large pale bill, and
legs pinkish.
Similar species None.
Voice Call loud *thip*.
Habitat and behaviour On
margins of mixed warbler
flocks, preferring low elevations
in the understorey of mangrove
and littoral and inland dry
forest and shrubland, and
Logwood on Grand Cayman.
Spreads tail.
Range Breeds in eastern

Adult. Cinnamon crown with black lateral stripes and black stripe through eye are diagnostic, March.

North America; winters on the
Caribbean slope from central
Mexico to Panama and, in
lower numbers, in the West
Indies: in the Bahamas, Greater
Antilles and Cayman Islands;
vagrant to the Lesser Antilles.
Status Fairly common, winter
visitor, August–April, although
hard to see; majority arrive in
small flocks and often leave by
March.

Louisiana Waterthrush
Parkesia motacilla

Taxonomy Monotypic.
Description L 14.5–16cm
(5.5–6.25in). Adult head and
upperparts dark olive-brown;
broad long supercilium buff in
front of eye, white and wider
behind, and extending to nape;
dark eye-line, white throat
unspotted, white underparts
with brown streaks on breast,
sides and flanks, flanks washed
bright buffy-yellow. No wing-
bars or tail spots.
Similar species Northern
Waterthrush smaller, supercilium
buff or whitish and tapers rather
than broadens behind eye.
Voice Loud *chink* lower pitch
than Northern Waterthrush.
Habitat and behaviour Mainly
terrestrial, often in more open
habitats than Northern, foraging

Adult. Note long supercilium becoming broader behind eye, December.

at wetland edges near lagoons
and ponds; near moving water
(e.g., where canals enter North
Sound on Grand Cayman).
Moves tail up and down and
from side to side continually and
in slow motion.

Range Breeds in eastern North
America; winters Florida to
Panama, and in winter and on
passage in the Bahamas, Greater
Antilles and Cayman Islands.
Status Rare passage migrant
and winter visitor, late July–April.

Northern Waterthrush
Parkesia noveboracensis

Adult. Buffy supercilium tapers behind eye towards nape, January.

Taxonomy Monotypic.
Description L 12.5–15cm (5–5.75in). Similar to Louisiana Waterthrush. Head and upperparts dark olive-brown, broad long supercilium is buff not bicoloured and narrows towards nape; underparts more uniform buffy or straw yellow with dark streaks on throat (fine), breast (heavy) and flanks; bill shorter, legs grey-pink.
Similar species Louisiana Waterthrush is larger, white throat is unspotted, supercilium is buff in front of eye and white and broad behind, white underparts.
Voice Loud metallic *pink*.
Habitat and behaviour Mainly terrestrial; selects understorey of red mangrove forest and shrubland around lagoons and herbaceous wetlands; arriving migrants occur in littoral/urban areas. Holds winter territories, constantly bobs head and pumps rump and tail rapidly up and down while walking in leaf litter or on low branches searching for arthropods and molluscs.
Range Breeds in northern North America; winters in southern Florida, Middle America to northern South America and the West Indies: mainly in the Bahamas, Greater Antilles and Cayman Islands.
Status Fairly common winter visitor and locally common passage migrant, late July–early June.

Adult has fine streaks on throat, seldom present in Louisiana Waterthrush, April.

Blue-winged Warbler
Vermivora cyanoptera

Taxonomy Monotypic.
Description L 12cm (4.75in).
Non-breeding male has nape
and mantle greenish, wings
blue-grey with two white wing-
bars, short black eye-line; yellow
crown, head and underparts
except white undertail-coverts;
slender pointed bill, and white
tail spots on outer rectrices.
Breeding male is brighter with
crown and underparts golden-
yellow, yellowish-olive back
and rump; female has crown
greenish-olive, eye-line greyish,
yellow forehead and duller
yellow underparts; first year
male similar. First year female
has upperparts olive-yellow,
narrow yellowish wing-bars,
bill pale.
Similar species Prothonotary
Warbler lacks wing-bars and
black eye-line.
Voice Silent.
Habitat and behaviour Edges

Adult male in breeding plumage, May.

of mangrove and dry forest,
and shrubland understorey.
Flicks tail.
Range Breeds in eastern North
America south to Georgia;
winters in Middle America,
rarely to Panama, and migrates
mainly through the Caribbean
Gulf slopes of Mexico. In the

West Indies, an uncommon
but regular passage migrant
and rare winter visitor in the
Bahamas, Cuba and Hispaniola,
and Cayman Islands.
Status Uncommon but regular
on passage, and rare in
winter, on Grand Cayman, late
August–late March.

Black-and-white Warbler
Mniotilta varia

Taxonomy Monotypic.
Description L 12.5–14cm
(5–5.5in). Non-breeding
male has has black and white
striped crown, face and
streaked upperparts; wide
white supercilium, two white
wing-bars, small tail spots
on outer rectrices, undertail-
coverts white with black spots,
variable amounts of white on
throat and underparts, bold
black streaks on sides and
flanks and blackish ear-coverts;
female duller, grey ear-coverts
and buffy flanks with blurred
narrower streaking. Breeding

Adult male in transition to breeding plumage, January.

Adult female non-breeding has grey cheeks, black post-ocular line and buffy flanks, April.

male is brighter with black throat; female more white than black, ear-coverts buffy-grey, white throat, buffy-white underparts with greyish streaking on sides and buffy flanks. First-winter resembles female but underparts more buff and streaking grey.

Similar species None.

Voice Usually silent, occasionally fast *chip-chip* call.

Habitat and behaviour Creeps around long branches and up and down the boles of trees foraging for arthropods; selects mangrove and dry forest (especially males who hold winter territories), also in woodland and dry shrubland, and urban/littoral areas on migration.

Range Breeds in North America; winters in Florida and the Gulf coast to Middle America and Colombia, Venezuela and eastern Ecuador, and the West Indies: where it is common in the Bahamas, Greater Antilles, Cayman Islands and northern Lesser Antilles.

Status Fairly common winter visitor and locally common passage migrant, July–May. Spring migration is noticeable in years when winter visitors have left by early April: a high proportion are males.

First-year male has white throat and broad white stripes on upperparts and white underparts, January.

Prothonotary Warbler
Protonotaria citrea

Adult male transitioning to breeding plumage, April.

Taxonomy Monotypic.
Description L 13.5cm (5.25in).
Adult male has golden head,
neck and underparts and
white undertail-coverts,
greenish mantle, wings and
tail blue-grey with extensive
white tail spots on rectrices,
black iris, long pointed black
bill paler at base in winter. No
wing-bars. Yellow can become
orange-yellow on breeding
male. Female duller; crown,
ear-coverts and nape greenish-
yellow, underparts dull yellow,
smaller tail spots. First-winter
male similar to female.
Similar species Blue-winged
Warbler has two white wing-
bars and black eye-line.
Voice Loud *chip*.
Habitat and behaviour Edges
of dry and mangrove forest,
and urban/littoral areas.
Range Breeds in eastern
North America to the Gulf
Coast and central Florida;
winters from the Yucatan
Peninsula to northern South
America, and rarely in Puerto
Rico eastwards. Migrates across
the Gulf where it is uncommon
on passage in the Bahamas,
western Greater Antilles (Cuba,
Hispaniola and Jamaica) and
Cayman Islands.
Status Uncommon passage
migrant, August–September,
and March–April, and rare in
winter.

Adult female, February.

Swainson's Warbler
Limnothlypis swainsonii

Adult bill is long, wide at base with straight culmen, April.

Taxonomy Monotypic.
Description L 14cm (5.5in).
Adults have brownish-olive back, wings and tail, flattened red-brown crown and nape, long wide whitish supercilium, long dark eye-line, yellowish-white underparts with brownish wash on flanks, long heavy pale bill wider at base, and pale pinkish legs. No wing-bars or tail spots.
Similar species Worm-eating Warbler has crown stripes.
Voice Silent.
Habitat and behaviour Forages in leaf litter in dry forest understorey; very secretive and inactive.
Range Breeds in the south-eastern United States to northern Florida; winters in the Yucatan Peninsula and, uncommonly, in the West Indies in the Bahamas and Greater Antilles west of Hispaniola; on passage in the Cayman Islands and Swan Islands.
Status Rare passage migrant, September–December.

Tennessee Warbler
Oreothlypis peregrina

Adult non-breeding, October.

Taxonomy Monotypic.
Description L 12cm (4.75in).
Non-breeding adults and breeding female have dull olive-green upperparts, yellowish-white supercilium, short dark eye-line, fine straight bill, yellowish wash on breast and abdomen, darker flanks, white undertail-coverts, short tail; adults have two faint wing-bars, no tail spots. First-winter has yellowish wash on undertail-coverts. Breeding male has grass-green mantle; blue-grey crown, nape and ear-coverts, white supercilium and whitish underparts.
Similar species Orange-crowned Warbler has yellowish lower eye-ring, faintly streaked yellowish underparts and yellow undertail-coverts.
Voice Call *tssst*.
Habitat and behaviour Forages in mangrove and dry forest, Logwood on Grand Cayman, gardens and littoral areas.
Range Breeds in northern North America; winters from southern Mexico to northern South America, a small number in winter in the West Indies.
Status Uncommon spring passage migrant and uncommon to locally common in autumn, rare in winter, September–April.

Kentucky Warbler
Geothlypis formosa

Taxonomy Monotypic.
Description L 12.5–14.5cm
(5–5.75in). All plumages have
dark crown, greenish-olive
upperparts, yellow underparts
and undertail-coverts, dark
iris, bill pale at base, legs
pinkish. Male has forehead
and forecrown blackish, black
from lores to below eyes
and on sides of neck, yellow
supercilium curves behind eye
forming partial 'spectacles'. No
wing-bars or tail spots. Female
and first-winter male have
reduced black markings on
crown and sides of neck. First-
winter female has olive crown
and flanks; greyish ear-coverts
and sides of neck.
Similar species Immature
Common Yellowthroat lacks
spectacles and has dusky brown
underparts. Canada Warbler
has grey upperparts, streaks
on breast and white eye-ring.

Adult male has extensive black below eye extending to sides of neck. Yellow
supercilium curves below eye in both sexes.

Hooded Warbler has yellow
face, and shows white outer
sides to tail in flight and when
tail is fanned.
Voice Usually silent, call *tik*.
Habitat and behaviour Mainly
terrestrial, in littoral and inland
dry forest and tall shrubland,
and mangrove forest edge;
arriving migrants may appear in
urban areas.

Range Breeds in south-eastern
North America (except Florida
and the Gulf coast); winters
mainly in Middle America (casual
in northern South America),
rarely in the West Indies: in the
Bahamas, Greater Antilles and
Cayman Islands, where it is
uncommon on passage.
Status Uncommon fall passage
migrant, August–December.

Common Yellowthroat
Geothlypis trichas

Taxonomy Polytypic (13).
Description L 11.5–14cm
(4.5–5.5in). Adult plumages
have brownish-olive upperparts,
no wing-bars or tail spots;
throat and undertail-coverts
yellow; sides and flanks washed
brownish. Breeding male has
wide black mask from mid-
crown, across eyes down to
sides of neck, bordered above
by ashy grey band; throat and
breast bright yellow. Female
duller, lacks mask, faint whitish
eye-ring, brownish ear-coverts,
less bright yellow throat and
breast. First-winter male has

Adult male, March.

Spring first-year male transitioning to adult plumage, April.

some blackish on face; throat and breast creamy yellow, underparts dusky brown; first-winter female very dull, pale eye-ring; buffy throat and breast and undertail-coverts, may have yellowish wash; brown on flanks.
Similar species Kentucky Warbler has yellow underparts in all plumages.
Voice Call grating *chack*.

Habitat and behaviour Low elevations in mangrove and Buttonwood shrubland around lagoons and herbaceous wetlands preferred; also in dry shrubland; arrives in small flocks in all habitats and disperses to become solitary.
Range Breeds throughout North America and northern Mexico; *G. t. trichas* winters from the extreme southern United States to Middle America (casually to northern South America) and the West Indies including the Bahamas, Greater Antilles and Cayman Islands; rare or vagrant in the Lesser Antilles. The endemic Bahama Yellowthroat *G. rostrata* is resident there.
Status Fairly common winter visitor and common passage migrant, August–May.

First-winter female, January. Adult female has faint supercilium and grey-brown ear-coverts.

First-winter female has pale eye-ring, buffy-yellow throat and undertail-coverts, January.

Hooded Warbler
Setophaga citrina

Taxonomy Monotypic.
Description L 12.5–14.5cm
(5–5.75 in). Male has black
hood (crown, nape and throat)
enclosing golden yellow
forehead and face, greenish
upperparts, bright yellow
underparts, black iris and pink
legs, extensive white tail spots
on outer rectrices. Female
usually has reduced hood with
blackish-brown cowl framing
face, yellow throat and dark
lores. No wing-bars. First year
male similar to adult male with
olive-yellow edges to hood;
first year female has olive-green
crown, nape, and upperparts;
face and underparts yellowish.
Similar species Wilson's
Warbler is smaller with no
tail spots. Kentucky Warbler
has blackish ear coverts in all
plumages, 'yellow spectacles'
and no tail spots.
Voice Usually silent, call
emphatic ringing *chip*.
Habitat and behaviour Forages
at low levels in littoral and
inland dry forest and shrubland,

Adult male, April.

also mangrove edge. Constantly
fans tail when perched.
Range Breeds in south-
eastern North America and
northern Florida; winters on
the Caribbean slope of Middle
America (mainly in south-
eastern Mexico) to Panama,
and the West Indies, where it

is very uncommon in winter
and uncommon on passage in
the Bahamas, Greater Antilles
(on passage only in Jamaica),
Cayman Islands and islands in
the western Caribbean.
Status Uncommon passage
migrant and winter visitor,
August– March.

Adult female.

American Redstart
Setophaga ruticilla

Taxonomy Monotypic.
Description L 11–13.5cm (4.25–5.25in). Male has black head, upperparts, throat and breast; brilliant orange patches on wings (base of primaries in flight), sides of breast and base of outer tail; white abdomen and undertail-coverts. Female has grey head, olive-grey upperparts, whitish supercilium, eye-ring and lores; orange patches replaced by lemon-yellow. First-year male resembles adult female; in first summer black appears on lores, head and breast, and yellow deepens to orange. First-winter female has olive-grey head and upperparts, pale yellow on sides and tail, wings often lack yellow.
Similar species None.
Voice Call soft high *chit.*
Habitat and behaviour Very active when foraging, fans tail, takes mainly insects by hover-gleaning and aerial fly-catching. Selects mangrove and dry forest, where males hold winter territories; also in littoral and urban areas, woodland and dry shrubland.

Adult male, March.

Range Breeds in North America including northern Florida and the Gulf states; winters in central Florida, and from southern Mexico to South America and, commonly, in the northern West Indies. It is the fourth most common migrant in Neotropical collections.
Status Common winter visitor and passage migrant, occurs August–early June. Late spring arrivals on passage are almost all males.

First-summer female, September.

First-summer male; note black lores and black spots on plumage, June.

Cape May Warbler
Setophaga tigrina

Adult male transitioning to breeding plumage, January.

Taxonomy Monotypic.
Description L 12.5–14cm (5–5.5in). Non-breeding male has greyish-olive back streaked blackish, rufous wash on ear coverts, yellow supercilium and patch on sides of neck; yellow throat and breast heavily streaked, white undertail-coverts, large white wing patch on median coverts, greenish rump, bill thin and slightly decurved, white tail spots on outer three rectrices. Female duller, two uneven white wing-bars, greyish ear-coverts, pale yellowish on sides of neck, thin streaking on breast. First-winter male has yellow supercilium and cheek patch, lacks chestnut; first-winter female is drab with greyish-brown upperparts; yellow

Adult male in breeding plumage, March.

Non-breeding adult female, November.

mostly absent, with blurred olive streaking on buffy-grey underparts. Breeding male has bright chestnut ear-coverts patch bordered by yellow sides of neck (almost forming a collar), orange-yellow supercilium, blackish crown, back heavily streaked black, and bright yellow rump; female has faint wing-bars, greyish ear–coverts and pale yellow on face, yellow underparts (rump greenish-yellow) with narrower streaking.

Similar species Magnolia Warbler has black-tipped tail with upper third white when seen from below, lacks yellow area on sides of neck; yellow underparts of immature are usually unstreaked. Yellow-rumped Warbler has yellow patches on sides and a brighter yellow rump in all plumages than all except unmistakable breeding male Cape May Warbler. Prairie and Palm Warblers have yellow undertail coverts and wag tail.

Voice Call high pitched *seet*.

Habitat and behaviour Gleans for arthropods from the forest canopy to near ground level in all habitats including urban/littoral areas, secondary growth and Logwood on Grand Cayman; takes fruits and berries, nectar and uses sapsucker wells on Little Cayman.

Range Breeds in North America; winters casually in Florida and Middle America, and mainly in the West Indies, where it is common in the Bahamas, Turks and Caicos Islands, Greater Antilles, Cayman Islands and uncommon in the Lesser Antilles.

Status Fairly common in winter, locally common on passage, September–late April. Populations fluctuate widely from year to year.

First-winter female, November.

Northern Parula
Setophaga americana

Taxonomy Monotypic.
Description L 10.5–12cm (4–4.5in). Adult plumages have blue-grey head, wings and tail; greenish mantle, two broad white wing-bar patches, two white eye crescents, bill with black culmen and yellow lower mandible continuous with yellow throat and breast; rest of underparts white, short tail with small white tail spots, and yellow feet. Breeding male has brighter plumage, blackish lores, and two bands across golden yellow breast forming a necklace (upper band blue-grey and lower chestnut); these indistinct in non-breeding male in winter; female has grey lores and rufous breast-band pale, a faint wash or absent. First-winter male resembles adult female, first-winter female has brownish-green wash on upperparts and less yellow on unmarked breast.
Similar species None.
Voice Call sharp *chip*; song short rising trill ending suddenly, heard occasionally in spring.

Adult male, February.

Habitat and behaviour Gleans for insects from the canopy to near the ground; also takes fruits, in forest and woodland habitats, littoral/urban areas, and Logwood/Buttonwood on Grand Cayman.
Range Breeds in eastern North America south to northern Florida; winters in Florida, Middle America to Panama, and mainly in the West Indies where it is common in the Bahamas, Greater Antilles, Cayman Islands and northern Lesser Antilles and uncommon in the southern Lesser Antilles, being the second most common migrant in the West Indies.
Status Fairly common to locally common winter visitor, August–late April.

Male non-breeding, January.

Adult female, December.

Magnolia Warbler
Setophaga magnolia

Taxonomy Monotypic.
Description L 11.5–12.5cm
(4.5–5in). Non-breeding adult
has grey head, white eye-ring,
narrow white wing-bars,
unstreaked bright yellow
throat; bright yellow breast and
abdomen with black streaks
on sides and flanks, heaviest
on male, fainter and indistinct
on female; yellow rump, white
undertail-coverts; underside
of tail with white band across
upper third is diagnostic, in
flight shows as white spots

Adult male breeding, April. Note diagnostic white band across underside of tail.

forming almost complete
subterminal band with black
terminal band. First-winter
has white eye-ring, unstreaked
greenish upperparts, and yellow
underparts with pale grey band
across upper breast. Breeding
adults have white supercilium
behind eye; male has black
cheek and back, white wing-
coverts patch, black breast-band
joining broad black streaks on
sides and flanks; female has two
white wing-bars, grey on cheeks
and back, and finer streaking on
yellow breast.

Similar species White band
under tail and tail pattern in
flight are diagnostic. Yellow-
rumped Warbler has yellow on
sides only, rest of underparts
white. Non-breeding male Cape
May Warbler has yellow neck
patch and densely streaked
yellowish underparts, including
throat.
Voice Call slurred repeated *tzek*.
Habitat and behaviour Edge
of mangrove and dry forest
and shrubland, urban/littoral
areas, and Logwood on Grand

Cayman. Gleans arthropods in
the outer branches of trees at
low to mid-levels.
Range Breeds in North
America; winters mainly in
Middle America, some south
to Panama; rare in winter and
uncommon on passage in the
West Indies in the Bahamas,
Cuba, Hispaniola, Jamaica and
Cayman Islands; occasional in
the Lesser Antilles.
Status Rare in winter, and un-
common on passage, August–
October and March–May.

Adult female non-breeding.

First-winter immature has pale band
across upper breast, November.

Bay-breasted Warbler
Setophaga castanea

Taxonomy Monotypic.
Description L 12.5–15cm (5–5.75in). All plumages have two broad white wing-bars, buffy undertail-coverts and extensive white tail spots. Adult females and first-year plumages have split eye-ring. Non-breeding adult has greenish-olive crown and upperparts with blackish streaks on mantle, short pale supercilium, grey eye-line, whitish throat and un-streaked underparts, chestnut wash on sides; flanks yellowish in first-winter. Breeding male has black forehead and face; dark chestnut crown, chin, throat, sides and flanks; heavily streaked back, cream patch on side of neck to hind-crown, rest of underparts whitish; breeding female lacks black mask, chestnut wash on crown and sides of breast only. First-winter male similar to adult; first-winter female has olive upperparts and buffy-whitish underparts including flanks. **Similar species** Non-breeding and immature Blackpoll

Adult male breeding plumage, April.

Warbler has white undertail coverts, faint streaking on sides, pale legs and yellow feet.
Voice Silent.
Habitat and behaviour Mangrove and dry forest, woodland, littoral/urban areas, especially in Casuarina trees along beaches.
Range Breeds in North America; winters from Panama to Colombia and north-western Venezuela. Migrates primarily across the Gulf in spring and on the Atlantic route in fall, with low numbers on passage in the West Indies in the Bahamas, western Greater Antilles and Cayman Islands and vagrant elsewhere in the region.
Status Uncommon but regular passage migrant, mainly in spring, October–November and April–late May.

Adult male in transition to breeding plumage, February. Note split eye-ring.

Adult female in transition to breeding plumage.

Blackburnian Warbler
Setophaga fusca

Taxonomy Monotypic.

Description L 13cm (5in). Non-breeding male has upperparts brownish-grey with buffy-white stripes on back, two white wing-bars, dark ear-coverts, various amounts of orange-yellow on crown, supercilium, neck patch, chin, throat and breast; blackish streaks on sides and flanks, whitish-buff underparts, triangular ear-covert patches and white undertail-coverts in all plumages. Female similar but upperparts olive, paler yellow on head and underparts, olive streaks on sides; has two white wing-bars in all plumages. Breeding male orange on face, throat and breast, black hind-crown, eye-line and ear-coverts, broad white patch on wing coverts, extensive white on outer rectrices, black streaking on sides of throat and breast to flanks, rest of underparts white; female has grey ear-coverts, yellow replaces orange on head and breast including

Male transitioning to orange breeding plumage on face and throat, April.

underparts, with less streaking. First-winter male resembles adult female but with black eye-line, yellower throat and heavier streaking on back and flanks; first-winter female very pale, face and throat buffy-white, faint streaks on sides and flanks.

Similar species Yellow-throated Warbler has white supercilium and unmarked grey back. Female and immature Black-throated Green Warbler have unstreaked greenish back and white chin and throat; immature also has white breast.

Voice Silent.

Habitat and behaviour In tree canopy in littoral/urban areas (often in Casuarina and Coconut Palms) and mangrove and dry forest.

Range Breeds in North America; winters from southern Middle America to north-western South America. Fall passage is mainly along the Atlantic coast and Florida, and the spring passage is mainly trans-Gulf; uncommon in western Cuba, Cayman Islands and islands in the western Caribbean and rare in the Bahamas and rest of the Greater Antilles.

Status Uncommon passage migrant, August–November and February–May, with the majority in late spring.

Adult female non-breeding.

Yellow Warbler
Setophaga petechia

Local name Yellow Bird.
Taxonomy Polytypic (43).
Description L 11.5–13.5cm
(4.5–5.25in). Adults are
yellow overall. Non-breeding
adult has yellowish-olive
upperparts, yellowish rump,
blackish-brown wings edged
bright olive-yellow, two yellow
wing-bars, yellow underparts
and undertail-coverts and
extensive yellow tail-spots
on outer rectrices, black iris,
slender black bill and grey
pink legs. Non-breeding male
has bright yellow head, some
cinnamon-rufous streaking on
breast and sides; breeding male
has saffron-yellow on crown
and bright rufous streaking
on breast and flanks. Breeding

Adult male showing yellow wing-bars and wing feathers edged olive-yellow,
March.

Adult male has rufous streaking on breast and yellow undertail-coverts, July.

female has olive green crown,
faint yellow eye-ring, olive wash
on sides, yellow breast unstreaked or with faint streaks.
Juvenile has greyish crown and
hindneck, whitish eye-ring and
throat: male has olive back and
yellowish underparts; female
has greyish back and face and
whitish-grey underparts.
Similar species Yellow tail spots
are diagnostic as are bicoloured
flight feathers. Prothonotary
Warbler has white undertail-
coverts and no wing-bars.
Hooded Warbler has white tail
spots and no wing-bars. Female
and immature Wilson's Warbler
(vagrant) are smaller, lack
wing-bars and tail spots, and
upperparts are greenish.
Voice Song loud, emphatic
clear *tseeet tseeet tseeet
ze-ze-ze tsweet*; call *chip,* also
zee-zee-zee.
Habitat and behaviour Coastal

Adult female. Note unstreaked underparts, May.

mangrove shrubland (including the edge of marine sounds) and forest edge (residents and migrants), Buttonwood around wetlands, uncommon in dry forest (on Cayman Brac breeds there irregularly); in gardens and littoral dry shrubland and woodland in migration. Gleans for arthropods and takes fruits. Pair bonds remain throughout most of the year and hold territories. Breeds, March–July, 2 eggs usual, bluish-white spotted with brown, in deep cup nest of grasses or Turtle Grass, spiders' webs, lined with palm fibre; female incubates and both parents feed young which fledge in around 14 days.

Range Breeds in North, Middle and northern South America. Fourteen endemic subspecies are resident in the West Indies; non-breeder in smallest Lesser Antillean islands. *D. p. eoa* breeds in the Cayman Islands and Jamaica. North American populations are migratory (*aestiva* group), observed in late winter and spring passage, although no migrants were reported in a study of 1,814 collected specimens from the West Indies.

Status Common to locally very common resident on Grand Cayman, and uncommon to locally fairly common on Little Cayman and Cayman Brac; migrants occur on passage.

First-year. Note grey head and whitish throat, October.

Chestnut-sided Warbler
Setophaga pensylvanica

Male in breeding plumage, May.

Taxonomy Monotypic.
Description L 11.5–13.5cm (4.5–5.25in). Non-breeding adult and first-winter have bright yellow-green crown, back and rump; faint streaks on back, two yellowish wing-bars, white tail spots on outer rectrices, prominent white eye-ring, grey face and neck, chestnut sides (absent on immature female, restricted in adult female) on whitish-grey underparts, and white undertail-coverts in all plumages. Breeding male has bright yellow crown, black eye-line to ear coverts and wide malar stripe; white cheeks, sides of neck, throat and underparts, wide chestnut band on sides and flanks, heavy streaks on back; breeding female duller with grey on face and chestnut on sides only.
Similar species Non-breeding Bay-breasted Warbler has split eye-ring and buffy undertail-coverts.

Voice Usually silent; call a loud *tsik*.
Habitat and behaviour Gleans arthropods at mid- to low elevations usually in edges of dry forest and shrubland, disturbed areas of secondary growth, rough pasture, and gardens. Holds tail cocked.
Range Breeds in eastern North America; winters from southern Mexico to Panama; on passage through the West Indies where it is uncommon in the western Greater Antilles in Cuba and the Cayman Islands and rare in the Bahamas and rest of Greater Antilles.
Status Uncommon passage migrant, September–November and February–May, most frequent in April and May.

Adult male non-breeding; female has little or no chestnut on sides.

Blackpoll Warbler
Setophaga striata

Adult male in breeding plumage showing tail spots, April.

Taxonomy Monotypic.
Description L 14cm (5.5in).
All plumages have mantle
heavily streaked black, two
white wing-bars, white tail
spots, white undertail-coverts
and feet yellow. Female has
split eye-ring. Non-breeding
adult and first-winter have
greenish-olive upperparts,
throat and breast yellowish-
buff with faint streaking on
sides and flanks, yellowish
supercilium, olive eye-line (male
retains narrow malar stripe).
Breeding male has black cap
to nape and black malar stripe,
white cheeks, white underparts
with black streaks from neck
to flanks; breeding female has
crown and upperparts grey
with black streaks, dark eye-
line, light streaking on whitish
underparts.
Similar species Immature
Bay-breasted Warbler has
buffy undertail-coverts and
unstreaked underparts, dark
legs and feet. Immature
Pine Warbler (Vagrant) has
unstreaked back, white eye-ring

and dark legs and feet. Black-
and-white Warbler has black
and white striped upperparts
and crown in all plumages.
Voice Usually silent, call clear
chip.
Habitat and behaviour Gleans
arthropods in the canopy of
dry and mangrove forest and in
littoral/urban areas.
Range Breeds in northern
North America; winters
mainly east of the Andes from
Colombia to Peru and Brazil.

Migrates almost exclusively
through the West Indies, in fall
along the Atlantic flyway from
Puerto Rico eastwards, and in
spring, through the Bahamas,
the western Greater Antilles:
where it is regular in Cuba, and
the Cayman Islands.
Status Uncommon to locally
fairly common passage
migrant, August–November
and April–late May. Most occur
in spring in small flocks that
linger briefly.

Adult female in breeding plumage, April.

Black-throated Blue Warbler
Setophaga caerulescens

Adult male, May.

Taxonomy Polytypic (2).
Description L 12–14cm
(4.75–5.5in). All plumages
show a diagnostic small white
rectangle on centre edge
of closed wing: reduced or
absent in immature female;
white undertail-coverts, and
small white tail spots on outer
rectrices, no wing-bars. Male
has dark blue upperparts; black
cheeks, throat and flanks, white
underparts. Female and first-
winter have upperparts olive-
brown, pale supercilium, white
lower eye-crescent in grey face,
and buffy-grey underparts.
Similar species No other
warbler has white primary
patch on closed wing.
Voice Call emphatic low *tik*.
Habitat and behaviour
Solitary, with males holding
winter territories in dry and
mangrove forest; females more
frequent in woodland and dry
shrubland, also in secondary
growth and littoral/urban areas.

Gleans arthropods at all levels
and takes fruits, berries and
nectar.
Range Breeds in eastern North
America; casual in winter in
southern Florida and Middle
America; most winter in the
West Indies: in the Bahamas,
Greater Antilles and Cayman

Islands, where both subspecies,
S. c. caerulescens and *S. c.
cairnsi*, occur; casual in the
Lesser Antilles.
Status Fairly common to
locally common winter visitor,
September–May (majority leave
by late April). One of the most
frequently seen warblers.

Adult female. Only female warbler with white square on wing, April.

Palm Warbler
Setophaga palmarum

Taxonomy Polytypic (2).
Description L 12.5–14cm
(5–5.5in). Yellow undertail-
coverts in all plumages. Non-
breeding adults and immatures
have brownish crown and
upperparts, white tail spots
on two outer rectrices, long
pale supercilium, long dark
eye-line, greenish-yellow rump,
underparts greyish-buff with
faint grey streaks on breast,
sides and flanks. Breeding adult
has chestnut crown, black
malar stripe; bright yellow
supercilium, chin, throat and
upper breast; fine rufous streaks
on breast and on buffy flanks.
Similar species Only warbler
with yellow undertail-coverts
and terrestrial habits that bobs
tail.
Voice Call *chip* similar to Prairie
Warbler.
Habitat and behaviour Bobs
tail continually like a Prairie
Warbler. Mainly terrestrial;
forages for insects and seeds;
also in low bushes. Occurs

Adult transitioning to breeding plumage. Note chestnut crown, March.

in urban and open disturbed
edge habitats, in littoral areas
including beach ridge and
Ironshore and in woodland and
dry shrubland. Holds winter
territories in urban areas.
Range Breeds in North America.
S. p. palmarum winters sparsely
on the United States Gulf coast

and Middle America, and
mainly in Florida and the West
Indies in the Bahamas, Greater
Antilles and Cayman Islands,
where it is common from
Hispaniola westwards.
Status Common winter visitor
and abundant in fall passage,
September–late May.

Non-breeding adult has yellow rump in all plumages, October.

Yellow-rumped Warbler
Setophaga coronata

Male in breeding plumage, May.

Taxonomy Polytypic (5).
Description L 14cm (5.5in).
Yellow rump, white wing-
bars and white throat in all
plumages. Non-breeding male
has greyish-brown upperparts
with streaks on mantle; small
yellow patches on crown
and breast sides; whitish
supercilium and lower eye
crescent; variable greyish-
brown streaks on breast and
sides, whitish undertail-coverts
and extensive white on outer
rectrices; female duller with less
conspicuous yellow on crown
and breast sides. First-winter
similar but even duller: the
most nondescript immature
warbler; head and upperparts
brownish, streaking on
underparts blurred, yellow may
show on rump only. Breeding
plumage seldom observed:
male has black forecrown,
lores and ear coverts; blue-grey
back with black streaks; yellow
crown cap and breast sides,
white underparts with black

breast-band and bold streaking
on flanks; female similar but
face grey, blackish streaking on
breast and sides.
Similar species Magnolia
Warbler has yellow underparts.
Cape May Warbler has greenish
rump.
Voice Call sharp *chep*.

Habitat and behaviour
Forages for insects and takes a
many berries in dry forest and
shrubland, littoral/urban areas,
secondary mangrove and rough
pasture.
Range *S. c. coronata* breeds
in North America; winters in
south-eastern United States,

Adult female non-breeding, February.

First-winter male and female indistinguishable in field, with less (or no) yellow on sides than non-breeding adult female, February.

First-winter has streaked underparts, long tail and yellow rump, February.

Middle America to Panama, and the West Indies: where it is common in the Bahamas, Cuba, Jamaica and Cayman Islands, uncommon to rare elsewhere, and of very irregular abundance in the region. Four other subspecies winter in

United States and Middle America.
Status Irregular passage migrant and winter visitor, rare to locally common; usually a late arrival from late December–April; uncommon August–November.

Yellow-throated Warbler
Setophaga dominica

Taxonomy Polytypic (4).
Description L 13cm (5in). Adult has grey crown and upperparts with no streaking, two white wing-bars, long white supercilium (may have yellow supraloral patch) and white patch on sides of neck, white lower eye crescent, black on ear-coverts continues down sides of canary yellow throat and breast; rest of underparts white with broad black streaking on sides and flanks, extensive white on outer rectrices. Breeding male has deeper yellow throat.
Similar species Kentucky Warbler has supercilium and entire underparts yellow. Blackburnian Warbler has white streaks on mantle and yellow supercilium.
Voice Call loud musical *chip*.
Habitat and behaviour Forages mainly for insects and spiders in the canopy of dry and mangrove forest, Buttonwood; in littoral/urban areas often in Casuarina, Coconut Palms and vines around houses; also takes nectar from century plants and vines.
Range Three subspecies breed in south-eastern North America and two, *S. d. dominica* and *S. d.*

Adult, February.

albilora, are migratory. *S. d. flavescens* is resident in the northern Bahamas. *S. d. dominica* winters in the south-eastern United States, the Bahamas, Greater Antilles and the Cayman Islands; *S. d. albilora* winters in Middle America to Panama and may occur in passage in the Cayman Islands. One of the earliest warblers to arrive in the West Indies.
Status Fairly common winter visitor, late July–May.

Vitelline Warbler
Setophaga vitellina

Local name Chip Chip.
Taxonomy Polytypic (3). Two endemic subspecies.
Description L 13cm (5in). On Grand Cayman, *S. v. vitellina* male has crown and upperparts olive–green (brighter when breeding), median wing-coverts edged yellow and greater coverts olive-yellow forming two wing-bars (indistinct in faded plumages, well defined when breeding), wings and tail dusty olive. Sides of head, wide supercilium and area below eye bright yellow, with olive-green eye-line and moustachial stripe that curves below eye to olive-grey ear-coverts. Underparts including undertail-coverts yellow; sides and flanks washed olive and faintly streaked olive-green, streaks often barely visible in winter. Terminal inner webs of three outer rectrices in male are white (tail spots), less obvious than in Prairie Warbler; bill horn, legs and feet blackish.

Adult *S .v. vitellina* has darker face markings than *S. v. crawfordi* on Little Cayman and Cayman Brac, April.

Female is similar, sometimes identical, but often duller with less distinct face markings. On Little Cayman and Cayman Brac, *S. v. crawfordi* is larger and with brighter yellow face, and eye-line, moustachial stripe and ear coverts are less defined; underparts are chrome yellow with fainter streaking on sides and flanks (occasionally face marking and streaking is more pronounced). Older juvenile/first-year plumages: *crawfordi* has a greyer head and nape, paler underparts and less defined wing-bars. Recently fledged juvenile *vitellina* has more defined eye-line and curved malar stripe, greyish-brown upperparts, and brighter yellow breast than *crawfordi*.
Similar species Prairie Warbler smaller, chestnut streaks on mantle (male), undertail-coverts yellowish and paler, tail spots much larger; immature always has less yellow on face and underparts than Vitelline Warbler and streaking on sides and especially flanks more visible in all plumages.
Voice Call *tsee* or *see*. There are three song types. Song of both subspecies before and at onset of breeding season is 4 syllables (first two level, third

Adult, Grand Cayman subspecies, has olive-grey eye-line, moustachial stripe and streaks on sides, April.

Adult breeding, Grand Cayman *S. v. vitellina*, has smaller and fewer tail spots than Prairie Warbler, May.

Immature plumage, Grand Cayman, July. Note broken whitish eye-ring and lack of yellow on face.

rising and fourth falling), *S. v. crawfordi* also has 3-syllable song with a long descending third note, both subspecies have a territorial song heard throughout the year of five, occasionally six, syllables (first three rising, fourth rising to a pitch and falling, the fifth lower than the first syllables).

Habitat and behaviour

Inland dry shrubland and open woodland with Thatch Palms preferred; also bushes in rough pasture, forest edge understorey, Logwood on Grand Cayman, and urban areas on Little Cayman and Cayman Brac. It is marginal in dry forest, and absent from mangrove and littoral areas on Grand Cayman where it is replaced by the Yellow Warbler; occasional in littoral urban areas of Cayman Brac. Forages from the ground up to 18m; a very active gleaner, occasionally hover-gleaning, taking mainly arthropods: especially spiders; also nectar from bromeliads and Agave. Flicks and fans tail frequently. Pair bonds from December may last throughout the year; nest built by female

Fledgling juveniles, Grand Cayman, July. Note greyish-brown crown, pale buff supercilium, pale chin, throat and undertail-coverts, and yellow breast.

Adult, *S. v. crawfordi*, Little Cayman, April.

Adult, *S. v. crawfordi*, is brighter yellow with face markings less defined than Grand Cayman subspecies, April.

is a small deep cup woven of thatch palm webbing, grasses and cotton, lined with bromeliad down, spiders' webs or feathers placed 1–6 m up. Most common nest types are a 55–60cm diameter open cup woven: between three uprights of a bush or tree branches and often hidden by leaves and vines, and a nest totally concealed in Banana Orchid or between two bromeliads with usually the lower plant dead and the upper alive. Because nests are often so well hidden few clutches and broods have been observed. Clutch of 2 greyish-white eggs mottled with grey and brown, April–August; both adults incubate and feed young. Males, females and immatures sing throughout the year.

Range Restricted Range species. The Cayman Islands and Swan Islands.

Status Resident, endemic subspecies S. v. *vitellina* is fairly common to locally common on Grand Cayman west of Savannah, and S. v. *crawfordi* is common on Little Cayman and Cayman Brac. The status of S. v. *nelsoni*, which is confined to Greater Swan Island (area 8 km²) is unknown. Thus the two populations in the Cayman Islands, in an area of 270 km², occupy c. 97% of the global range. Conservation status: Near-threatened.

Adult female on nest, Little Cayman, May.

Juvenile/First-winter *S. v. crawfordi*, Cayman Brac. Note grey crown, nape and sides of neck, yellow on supercilium in front of eye and below eye, buffy throat and pale yellow breast streaked greyish, December.

Prairie Warbler
Setophaga discolor

Adult male, April.

Taxonomy Polytypic (2).
Description L 12cm (4.75in).
All plumages have black
or blackish spot in front of
wing. Male has olive-green
crown and upperparts, broad
yellow supercilium and yellow
crescent below eye, black
eye-line and black moustachial
stripe curving below eye,
olive-green ear coverts, bold
chestnut streaks on mantle,
bright yellow underparts with
heavy black streaking on sides
and flanks, whitish-yellow
undertail-coverts. Wings
slate-grey with two yellowish
wing-bars (may appear as
single yellow patch coverts or
second bar indistinct) and large
white tail spots on three outer
rectrices. Female is less bright,
chestnut on mantle reduced,
eye-line and moustachial stripe
dark olive, underparts yellow
with indistinct blackish-olive

First-winter male has blackish face markings, February.

First-winter female. Note eye-arcs, absent in adults. Similarity to Vitelline Warbler, except for blackish spot in front of wing, present in all plumages, April.

streaking, tail spots smaller. First-winter has two white eye arcs, male may resemble adult or face pattern less contrasting and streaks may be narrow and blackish; female dull with olive upperparts, greyish head and cheeks, whitish (not yellow) short supercilium and below eye, chestnut streaking absent, spot in front of wing grey, yellowish underparts with olive streaking on sides and flanks, always more pronounced than in Vitelline Warbler.

Similar species Vitelline Warbler (often confused with this species) is larger, lacks chestnut streaks on mantle and black spot on sides of neck, face markings are olive-green not black or blackish and less well defined; lacks olive wash on flanks or shows diffuse olive streaking, undertail-coverts bright yellow; immature Vitelline Warbler has yellower supercilium, cheeks

and underparts than immature Prairie Warbler. First year Magnolia Warbler has white eye-ring and yellowish rump.

Voice Call low *chip*.

Habitat and behaviour Prefers edge habitats of mangrove and dry forest and mangrove shrubland, littoral/urban areas, and Logwood/Buttonwood on Grand Cayman; infrequently in dry shrubland, the preferred habitat of the Vitelline Warbler. Forages at all elevations gleaning and hover-gleaning mainly for insects and spiders; also takes nectar (of Century Plants in the Sister Islands and mangrove flowers). Constantly bobs and wags tail.

Range Breeds in south-eastern North America to the Atlantic Gulf coast; *S. d. discolor* winters in central Florida, islands off the Caribbean slope of Middle America, and the West Indies: in the Bahamas and Greater Antilles and Cayman Islands.

Status Fairly common in winter and locally common fall migrant, August–June; the majority have left by mid-May.

First-winter. All plumages have more extensive tail spots than Vitelline Warbler, February.

Black-throated Green Warbler
Setophaga virens

Adult male, April.

Taxonomy Polytypic (2).
Description L 12.5cm (5 in). Non-breeding adult has crown and upperparts bright olive-green with no streaks on mantle, dark wings with two broad white wing-bars, face yellow with wide yellow supercilium to nape, dark eye-line with yellow crescent below eye, olive ear-coverts, white underparts with black streaks on sides and flanks, white undertail-coverts with yellow wash on vent, white tail spots on outer rectrices. Breeding male has black chin, throat, upper breast and sides and heavy streaks on flanks; female and first-winter male duller with whitish chin, reduced or mottled black on throat and breast retained on sides and flanks, male has black spots on throat. First-winter female has olive ear coverts, whitish underparts may be washed yellow with faint streaking on sides.
Similar species Blackburnian Warbler has yellow or orange throat and breast, whitish stripes on back.
Voice Call high-pitched weak *tsip*.
Habitat and behaviour Dry and mangrove shrubland preferred, also forest edges, secondary growth, littoral/ urban areas, and gardens; gleans insects and takes some fruits.

Range Breeds in North America; *S. v. virens* winters mainly from southern Florida through Middle America to Panama, and uncommonly to northern South America and the West Indies. *S. v. waynei* breeds in south-eastern United States.
Status Uncommon winter visitor and fairly common passage migrant, September–May.

Adult female transitioning to breeding plumage, April.

Canada Warbler
Cardellina canadensis

Taxonomy Monotypic.
Description L 12.5–15.5cm
(5–5.75in). Dark blue-grey
upperparts, dark iris with
wide white eye-ring, yellow
underparts and white undertail-
coverts, pale pinkish legs
and long tail. Male has black
forehead, forecrown and
sides of throat; black streaked
necklace across yellow breast.
No wing-bars or tail spots.
Female similar but duller, face
and breast necklace greyish
(not black) and less distinct.
First-winter similar with olive-
grey upperparts and necklace
faint or absent.
Similar species No other
warbler has comnination of
blue-grey upperparts, white
eye-ring and white undertail-
coverts.

Adult male in breeding plumage.

Voice Silent.
Habitat and behaviour
Forages in canopy of dry forest
and emergent trees in dry
shrubland, gleans and hover-
gleans for insects.
Range Breeds in North
America; winters in South
America from Colombia to Peru
and Brazil. On passage mainly
through Middle America, rarely
in the Bahamas and Cuba (fall)
and Cayman Islands; casual
or vagrant elsewhere in the
Greater Antilles.
Status Rare passage migrant,
September–November and
February.

Bananaquit
Coereba flaveola

Local name Bananabird.
Taxonomy Polytypic (41).
Description L 10–12.5cm
(4–5in). Adult has dull black
upperparts, yellow rump, wide
white supercilium, black eye-
line to nape, yellow shoulder,
red skin at base of sharply
pointed decurved black bill (red
gape), greyish-white throat,
abdomen and undertail-coverts;
bright yellow breast, white
triangle at edge of closed
wing, white tail spots on
short tail. Juvenile dull with
greyish-brown upperparts and
buffy-grey underparts with
yellow patches, greyish-yellow
supercilium.
Similar species No other

Adult has red spot on gape and white square on wing, December.

Juvenile, May.

landbird has a pointed decurved bill.

Voice Call *tsiip,* grating varied song including series *churr-rr-rr-rr,* and level high *te-zi-te-zi-te-zi.*

Habitat and behaviour In all habitats, except forest interior, with inland and littoral dry shrubland (and xeric shrubland)

Adult with young *c.* 9 days old, December.

preferred, also urban/littoral areas and gardens. Forages from canopy to close to the ground, feeding primarily on nectar (flower piercer) moving rapidly between flowers; also takes fruits and small insects. Breeds throughout the year if rains plentiful, peaks February–September. Several clutches of 2–3 olive eggs with reddish spots, in globular nest with entrance at lower side, loosely woven with Red Birch bark, grasses, vines, fibres and stripped palm, lined with Red Birch, usually at outer end of branch or protected in poisonous trees or trees armed with thorns: such as Lady Hair, Shake-Hand, Manchineel, Dildo cactus; also in vines around houses and even in nest boxes. Continual nest builder, nests also built for roosting.

Range Resident in Middle and South America and throughout the West Indies, excluding mainland Cuba but including Cayo Coco and cays off the north Cuban coast, and Swan Islands.

Status Endemic subspecies *C. f. sharpei* is a common to abundant resident.

Western Spindalis
Spindalis zena

Local name Bastard cock. Previously named Stripe-headed Tanager.
Taxonomy Polytypic (5).
Description L 15cm (5.75in). Male has greenish-yellow mantle; brownish-orange nuchal collar, rump and uppertail-coverts; black head with broad white supercilium and white sub-moustachial stripe, black malar stripe; short thick dark-tipped bill with dark grey upper mandible and silver grey lower mandible. White chin, orange-chestnut throat and upper breast (extent and density varies), yellow on lower breast tapers into the mid-abdomen; abdomen and undertail-coverts greyish; black wings edged white, broad white patch on coverts, tail black and outer rectrices edged

Adult male, Grand Cayman only, January.

Adult male, October.

white, legs grey. Female is dull, upperparts greyish with olive-green wash on ear coverts sides of neck, rump and uppertail-coverts; wings edged whitish or greenish with small square whitish primary patch on closed wing; whitish supercilium and sub-moustachial stripe, underparts greyish-white. Immature male has greyish head with whitish stripes, breast is yellowish with blackish streaks on back. Immature female and juvenile have primary patch usually absent.
Similar species None.
Voice Call *seep*; one song is slight, almost whispered warble of 8–10 *tswee-tswee* notes rising, increasing in volume, falling then ceasing abruptly; second song is loud almost

Adult female. Note white square on closed wing, Grand Cayman only, January.

parrot-like chattering.

Habitat and behaviour
Forages in all habitats, except coastal red mangrove, at all levels; for fruits and berries, also flowers and young leaves. Breeds in dry forest and shrubland, March–August, occasionally second brood to December; pair bonds are maintained throughout the year. Flimsy cup nest of grass and soft fibre above 3m in understorey; 2–3 creamy-white or whitish-blue eggs, spotted red; young fed by both parents.

Range *S. zena* is resident on Bahamas, Cuba, Cayman Islands and the island of Cozumel, Mexico; vagrant to Florida. There are five subspecies: *S. z. pretrei* on Cuba. *S. z. salvini* on the Cayman Islands, two subspecies in the Bahamas and one in Mexico.

Status Endemic subspecies *S. z. salvini* is a fairly common resident on Grand Cayman only.

Juvenile, October.

First-year female lacks white square on wing of adult, March.

Cuban Bullfinch
Melopyrrha nigra

Local name Black Sparrow.
Taxonomy Polytypic (2).
Description L 14–15cm
(5.5–5.75in). Male is black
above and below with greenish
gloss on upperparts, deep
heavy black bill with convex
culmen, black iris, closed
wing has white border; short
rounded tail. In flight short
rounded wings show white
on part of primary coverts,
inner webs of outer primaries
and secondaries, axillaries and
underwing-coverts. Female has
dull slate-black plumage tinged
with olive, darker on head and
paler on lower underparts,
bill silver-grey, wing shows
less white. Juvenile similar to
female with white on wing
reduced or absent, feathers
tipped greenish-olive, bill pale.
Immature male becomes black
on head and wings in first
spring.
Similar species None.
Voice Call insect-like *chi-p* and

Adult male. White band on wing shows as white primary coverts and
underwing-coverts in flight, April.

Adult male, April.

zee zee, the first note high
pitched; song begins as a trill
zee-zee-zee, falls briefly then
rises for between 8 to 30 *tssi*
notes, the longest reaching
very high and barely audible
pitch, only heard at onset of
breeding season.
Habitat and behaviour
Usually in small flocks: of
males or females or one
male with several females.
Forages for seeds and fruits
(insects in breeding season)
at all levels from the canopy
and understorey to near the
ground, in woodland and dry
shrubland, Buttonwood and
mangrove edge, rough pasture

Adult female, Grand Cayman only, March.

Immature female, March.

Juvenile male, August.

and inland gardens; seldom in littoral areas. Pairs have prolonged courtship; males display with wing-flashing from January exposing white axillaries. Both sexes construct several nests until egg laying begins, March–August, in preferred trees of Silver Thatch and Shake Hand. There are two nest types: a large, untidy enclosed globular nest with side entrance; of twigs, vines, grasses, hair and plant fibre, lined with Red Birch bark, often

Juvenile, August.

within a tangle of Vine Pear vines and Scorn-the-ground; or an open arch of twigs backed by the tree bole with a rough nesting cup. Clutch 3–4 dull white eggs with greenish wash and reddish-brown spots, females incubate and both adults feed young; nests also built for roosting.

Range Restricted Range species and Greater Antilles endemic to Cuba and the Cayman Islands. Endemic subspecies *M. n. nigra* breeds on Cuba and the Isle of Youth, and *M. n. taylori* breeds on Grand Cayman.

Status Fairly common to locally common resident endemic subspecies *M. n. taylori,* on Grand Cayman east of Savannah (common in the Botanic Park). Its range has contracted and it is rare to absent west of Savannah due to urban development and loss of habitat. Studies

Adult female alighting and displaying white underwing-coverts, January.

indicate significant differences in morphology and song from the Cuban subspecies and it is expected that the two will be accepted as separate endemic species.

Immature male feeding on Logwood flowers, February.

Male in nest in Shake Hand tree, Grand Cayman only, April.

Yellow-faced Grassquit
Tiaris olivaceus

Local Name Grass bird.
Taxonomy Polytypic (5).
Description L 11.5cm (4.5in).
Male has brownish-olive
upperparts; wide supercilium,
chin and throat rich yellow;
pale yellow curve along lower
eye-lid; blackish lores, malar
and upper breast; rest of
underparts greyish, bill small
and straight, legs grey. Female
has paler underparts, no
black on breast, pale yellow
supercilium and chin; juvenile is
darker than female with short
buffy supercilium and chin.
Similar species None.
Voice Song a varied series
of weak high pitched trills
*zee-zee-zeeeeeeee zee-zee-
zeeeeeee*; call *quit*.
Habitat and behaviour Often
in small flocks in dry shrubland,
disturbed habitats, rough
pasture, littoral/urban areas
and inland gardens; foraging

Adult male, May.

Adult female lacks black on breast, May.

for seeds and small fruits, usually on the ground. Breeds in Casuarina, Logwood/Buttonwood on Grand Cayman, urban/littoral areas, gardens and dry shrubland; nesting peaks February–July, but nests throughout the year on Grand Cayman. Domed nest with side entrance woven of grasses in fork of tree or shrub 0.5–6m up; clutch 2–3 whitish eggs with spotted and marked brown; nests also used as roosts.

Range Resident from Mexico to northern South America, and the West Indies, where *T. o. olivaceus* breeds in the western Greater Antilles including Hispaniola, and Cayman Islands.

Status Locally fairly common breeding resident on Grand Cayman, uncommon on Cayman Brac and inhabited areas of Little Cayman.

Juvenile has buffy supercilium and chin, October.

Chipping Sparrow
Spizella passerina

Taxonomy Polytypic (5).
Description L 12.5–14cm (5–5.5in). Non-breeding adult has reddish crown with white median spot in forecrown, yellowish-grey supercilium, black eye-stripe and lores, thin malar and sub-malar stripes, bill horn, cheeks grey-buff; upperparts brown with dark brown streaks, brown rump with fine streaking, long tail notched, legs flesh. Breeding adult has bright rufous crown, whitish supercilium, black bill, whitish throat, greyish-buff breast and grey rump. Immature similar but face, breast and flanks have blackish streaks.
Similar species Clay-colored Sparrow has white median crown stripe, pale lores, dark brown cheek patch outlined in black.

Adult in breeding plumage, May.

Voice Silent.
Habitat Forages for seeds on the ground in urban gardens with feeders.
Range Breeds in North America and Middle America. Migratory northern races winter in the south of the breeding range; rare in winter in the northern Bahamas and Cuba from October–April, vagrant elsewhere.
Status Vagrant, recorded on Grand Cayman, October–November.

Grasshopper Sparrow
Ammodramus savannarum

Taxonomy Polytypic (11).
Description L 12.5cm (5in).
Upperparts brownish-grey
striped with tawny and grey,
black centres to wing-coverts,
yellow on bend of wing, flat
head with whitish median
crown-stripe, prominent eye-
ring around dark iris in pale
buffy face, partial supercilium
orange-red in front of eye and
buffy behind, dark spot on ear
coverts, breast is unmarked
cinnamon-buff, abdomen
whitish. Large, horn coloured,
long bill; short tail with
pointed rectrices and pink legs.
Immature similar but breast
and flanks finely streaked.
Similar species Savannah
Sparrow has heavily streaked
breast and black malar stripe.
Clay-colored Sparrow (Appendix
6) and Chipping Sparrow have
buff supercilium, grey nape and
greyish underparts, lack spot on
supercilium.
Voice Call *kr-it*.
Habitat and behaviour On the
ground in rough pasture and

Adult has supercilium orange in front of eye, greyish behind, February.

littoral areas, usually hidden in
long grass.
Range Breeds in North America
including Florida; sedentary
populations in Middle America,
Panama, north-western South
America and the West Indies in
Jamaica, Hispaniola and Puerto
Rico. Two North American

subspecies winter in the United
States and Middle America;
of these *A. s. pratensis* also
winters in the West Indies in
the Bahamas and Cuba and
Cayman Islands.
Status Rare passage migrant on
Grand Cayman, September–May.
Formerly fairly common in winter.

Summer Tanager
Piranga rubra

Taxonomy Polytypic (2).
Description L 18–19.5cm
(7–7.5in). Adult male plumage,
retained in all seasons; entirely
rose-red with underparts
brighter, crown slightly crested,
large stout bill, darker red
wings and tail, legs grey, shows
yellow axillaries in flight. Female
has rich yellow head and
upperparts, whitish eye-ring,
yellowish-orange underparts.
Immature male similar to

Adult male, October.

Adult female, October.

female, becoming increasingly red on head and breast on first spring.

Similar species Scarlet Tanager female is greenish-yellow; non-breeding male similar but brighter green, blackish wings and tail, axillaries white not yellow, smaller shorter bill.

Voice Call loud, rapid *pit-i-tuk*.

Habitat and behaviour Littoral and littoral/urban areas and dry forest and shrubland; foraging on insects, fruits and seeds including Royal Palm.

Range Breeds in the United States and northern Mexico; winters from Middle America to South America; *P. r. rubra* is uncommon on passage and rare in winter in the West Indies in the Bahamas, Cuba, Jamaica and Cayman Islands; casual or vagrant from Hispaniola eastwards to the Lesser Antilles.

Status Rare in winter, uncommon to fairly common spring passage migrant, occasionally locally common in fall, arriving in flocks with thrushes and warblers. Recorded August–May.

First-year immature male, April.

First-year immature male, April.

Scarlet Tanager
Piranga olivacea

Adult male, May.

Taxonomy Monotypic.
Description L 18cm (7in). Female in all plumages has olive-green to greenish upperparts, yellowish underparts, dark olive wings and tail edged brownish; non-breeding male similar with brighter yellow underparts, blackish wings and tail with no olive edging, axillaries white. Breeding male (in late spring) is bright scarlet with black wings and tail, small stout bill, grey legs. Immature male is greenish-yellow becoming increasingly mottled with red.
Similar species Summer Tanager male is entirely red; female and immature male are more orange-yellow than greenish, wings and tail dusky, bill longer and heavier, axillaries yellow.
Voice Silent.
Habitat and behaviour Very inactive; in canopy of dry forest and trees in littoral/urban areas.
Range Breeds in North America; winters in South America, and occurs on passage in Middle America, Netherlands Antilles and the West Indies: where it is very uncommon to rare in winter and on passage in the Bahamas, Cuba, Jamaica and Cayman Islands (except during fall migration); absent from Hispaniola.
Status Uncommon to locally fairly common fall passage migrant, September–early November, uncommon in spring, February–late May, and rare in winter.

Adult male non-breeding resembles female except wings and tail are black, February.

Rose-breasted Grosbeak
Pheucticus ludovicianus

Taxonomy Monotypic.
Description L 19–20cm
(7.5–8in). Very large conical
horn-coloured bill. Non-
breeding male has brownish-
black head and upperparts with
white spots on wing coverts,
breast pink and underparts
buffy. Breeding male has glossy
black head, throat, upperparts
and tail; white rump, black
wings with two large white
patches on coverts, crimson
V on central breast; rest of
underparts white with streaking
on flanks. Breeding female has
dark brown striped upperparts,
buff central crown-stripe, small
whitish wing-bars, wide whitish
supercilium to behind ear
coverts, white crescent below
eye, white malar stripe and
throat; breast and flanks buffy-
white and heavily streaked.
First-winter similar to female
but male has pink wash on
upper breast, brown edgings to
blackish upperparts and greyish
rump. In flight male shows
white outer tail feathers, red
underwing-coverts with white
inner primaries, female has
yellow underwing-coverts.

Male in breeding plumage, April.

Similar species None.
Voice Call short, thin insect-like
chink.
Habitat and behaviour Mid-to
low levels when on passage in
urban and littoral/urban areas,
dry shrubland, and Logwood
on Grand Cayman; also in dry
forest in winter.
Range Breeds in North America.

Winters in Middle America to
northern South America, and
the West Indies where it is
uncommon; regular on passage
in the Bahamas, Greater Antilles
and Cayman Islands; vagrant to
the Lesser Antilles.
Status Uncommon winter
visitor and locally common
passage migrant, October–May.

Adult female, April.

First-summer male. Note brown primaries and whitish
edges to head and back feathers, April.

Blue Grosbeak
Passerina caerulea

Taxonomy Polytypic (7).
Description L 16.5–19cm
(6.5–7.5in). Large, heavy bill,
flattened crown and long
rounded tail. Non-breeding
male has feathers edged
brownish on body and whitish
on undertail-coverts. Breeding
male is bright ultramarine
blue, dark wings with tan
wing-bars (upper is larger in
both sexes), black lores and
chin, heavy conical bill with
blackish upper mandible and
silver lower mandible. Breeding
and non-breeding female have
grey-brown or rufous-brown
plumage, paler on face and
underparts, two buffy-tan
wing-bars, blue feathers often
visible on rump, wing-coverts
and tail. In first-winter both
sexes are bright brown with
rufous wing-bars; male
develops blue on head and
wings.
Similar species Indigo Bunting
has smaller bill: male lacks

Adult male in breeding plumage, April.

tan wing-bars, female and
immature have faintly streaked
breast and white throat.
Voice Call resonant *click*.
Habitat and behaviour Small
flocks usual, occasionally in
large flocks in migration;
feeding on the ground or in
bushes on grass seed and
arthropods in dry shrubland,
littoral/urban areas, rough
pasture; may be among flocks
of the more common Indigo
Bunting. Constantly flicks and
fans tail.
Range Breeds in the United
States and Middle America;
two northern subspecies are
migratory wintering from
Middle America to Panama, *P. c.
caerulea* is rare to uncommon
in winter and uncommon on
passage in the Bahamas, Cuba,
Hispaniola, Puerto Rico and
Cayman Islands; vagrant to
Jamaica.
Status Uncommon,
occasionally locally common,
passage migrant, September–
October and March–May;
majority observed on Grand
Cayman.

Adult female has massive bill and tan wing-bars, October.

Indigo Bunting
Passerina cyanea

Adult male in breeding plumage, April.

Taxonomy Monotypic.
Description L 14 cm (5.5in).
Small conical bill and short tail.
Non-breeding male appears
brown with blue feathers
partially covered by brownish
edges on head, wings and
breast. Non-breeding and
breeding female are cinnamon-
brown overall with white throat,
faint buffy wing-bars and light
streaking on breast; some have
blue tinge on wing-coverts,
edge of wing, rump and tail.
Breeding male is entirely blue
becoming indigo on head and
breast; in flight shows brownish-
grey flight and tail feathers.
First-year male is brown with
blue feathers emerging on body,
head, wings and tail.
Similar species Blue Grosbeak
has larger bill, male is deep
blue but with tan wing-bars.

Female in breeding plumage; note blue on point of wing, April.

Adult female showing white throat and streaked underparts, April.

Adult female, April.

Voice Males sing on late spring migration, long series of insect-like *swee swee sweer* and short trills or clicks; call *chip*.
Habitat and behaviour Usually in small flocks in open dry shrubland, littoral and inland urban areas, rough pasture and plantations; foraging for seeds, fruits, insects and spiders.
Range Breeds in North America; winters from Middle America to Panama, and the West Indies in the Bahamas, Greater Antilles and Cayman Islands, where it also occurs on passage; vagrant to the Lesser Antilles.
Status Fairly common winter visitor and common passage migrant, October–May.

First-year male, April.

Painted Bunting
Passerina ciris

Taxonomy Polytypic (4).
Description L 14cm (5.5in).
Male is unmistakable with a
purple-blue head, yellow-green
mantle; red rump, throat,
underparts and eye-ring.
Female and immature male
have pale eye-ring, olive-green
upperparts and olive-yellow
underparts.
Similar species None.
Voice Silent.
Habitat and behaviour All
observations have been at bird
feeders in littoral/urban areas
with Indigo Bunting, except for
one female in dry shrubland.
Range *P. c. ciris* breeds in
south-eastern United States
and winters in the Bahamas
and Cuba where now very
uncommon in winter and
uncommon on passage;
vagrant to Jamaica.
Status Rare passage migrant
and winter visitor, December–
April. A male, female and
immature remained for a
month in winter 1998.

Adult male, March.

Adult female breeding, April. Non-breeding and immature male have olive-green plumage.

Dickcissel
Spiza americana

Taxonomy Monotypic.
Description L 15–18cm
(5.75–7in). Non-breeding adult
has brownish-grey upperparts
with black streaking, rufous
coverts (shoulder), dark wings
edged with buff, greyish
crown and ear coverts, long
yellowish supercilium, black
malar stripe and broad pale
sub-moustachial stripe, white
chin and throat, grey conical
bill, and pointed central tail
feathers; non-breeding male
has pale but defined breast
patch. Breeding male has grey
nape, black V on throat and
central breast, yellow breast,
rest of underparts greyish-
yellow. Breeding female smaller
with yellowish-buff wash on
breast (black absent) and
indistinct streaking on sides
and flanks. First-winter is dull

Adult male in breeding plumage, April.

Adult female, April.

greyish-brown overall, rufous
coverts absent, back heavily
streaked and breast faintly
streaked.
Similar species None.
Voice Silent.
Habitat and behaviour Forages
for seeds in rough pasture
and open dry shrubland with
emergent trees.
Range Breeds in central
North America; winters in
Middle America to northern
South America, and occurs
on passage through Middle
America, rarely the West Indies
in the Bahamas, Cuba, Cayman
Islands, Providencia and San
Andres; no recent records from
Jamaica or Hispaniola.
Status Rare, but increasing,
passage migrant in spring,
April–May, occasionally
abundant in fall-out.

Bobolink
Dolichonyx oryzivorus

Taxonomy Monotypic.
Description L 18.5cm (7.25in).
Non-breeding adults have flat
black crown with buff nape
and central crown stripe,
black post-ocular stripe, buffy
throat, bright yellow-buff body
plumage and rump, black
stripes on back, two faint wing-
bars; black streaking on sides,
flanks and lower abdomen;
heavy pinkish bill and pointed
tail. Breeding male has pale
yellowish-buff hindcrown and
nape, rest of plumage black
except for whitish-grey rump,
white scapulars and gold streaks
on back, wing feathers edged
whitish-buff, dark bill. Breeding
female similar to non-breeding
adult with greyish-buff nape,
whitish throat, buffy-grey
underparts, fainter streaking on
flanks and grey rump.
Similar species All sparrows are

Adult male, March.

Adult female, April.

smaller and lack streaking on
sides and lower abdomen and
the pointed tail feathers.
Voice Loud *pink* repeated.
Habitat and behaviour Males
and females are often in separate
small flocks, foraging for seeds
and insects in rough pasture,
grassland, littoral shrubland, and
herbaceous wetlands.
Range Breeds in North America;
winters in central South America;
on passage through the West
Indies where it is most frequent
in the Bahamas, Cuba, Jamaica,
Cayman Islands, Providencia and
San Andres, and uncommon to
rare elsewhere.
Status Fairly common passage
migrant in small flocks,
September–early November, and
March–May. More regular in
spring on Little Cayman.

Greater Antillean Grackle
Quiscalus niger

Adult male, Grand Cayman subspecies, March.

Local name Ching-ching, Cling-cling.
Taxonomy Polytypic (7). Two endemic subspecies.
Description L 25–30cm (10–12in). Adults of both subspecies have yellow iris and long conical pointed bill. Male is large with glossy metallic violet-blue to blackish-violet body, wing coverts iridescent bronze-green, and long V-shaped bluish-green tail. Female duller, tail smaller with slight V. Juvenile dull brownish-black, tail normal shape, iris light brown. Little Cayman subspecies is smaller.
Similar species Female Shiny Cowbird is smaller, with shorter bill and dark iris.
Voice Call *ching ching ching;*

Adult male, Little Cayman subspecies, feeding on nectar of Century Plant, April.

Adult female, Grand Cayman subspecies, November.

musical 4-syllable fluting song; also sharp *cluck,* and raucous begging call of juvenile.

Habitat and behaviour
Gregarious, found in all habitats. Raids nests for young birds and takes fledglings (mainly Bananaquit, grassquit and ground-dove), snakes, frogs, arthropods, and also fruits; scavenges around humans at beach hotels; takes nectar of Century plants on Little Cayman. Both subspecies have large roosting and breeding colonies in mangrove and Buttonwood forest and shrubland, and smaller colonies and single pairs often in Royal Palm, Silver Thatch, Coconut Palm and exotic palms in urban/littoral areas. Pairing, prolonged courtship and nest building from February, and egg laying, March–August; on Little Cayman breeds April–July. Bulky cup nest of palms, red birch bark and grasses lined with soft plant material, built by both adults; 2–4 olive eggs with reddish markings, and young

fed by both parents.
Range Greater Antillean endemic. Two subspecies resident on Cuba and cays, two the Cayman Islands, one each on Jamaica, Hispaniola and islands, and Puerto Rico.
Status Endemic subspecies *Q.*

n. caymanensis is abundant on Grand Cayman. *Q. n. bangsi* is fairly common to locally common in the central and western parts on Little Cayman, becoming occasionally very uncommon in winter; it is extirpated from Cayman Brac.

Juvenile, Grand Cayman subspecies. Note dark iris, May.

Shiny Cowbird
Molothrus bonariensis

Taxonomy Polytypic (7).
Description L 18–20cm (7–8in). Has a dark iris in all plumages. Male is glossy black with purple iridescence and short, sharply pointed conical bill. Female has dull greyish-brown upperparts; pale buff supercilium and underparts. Juvenile similar to female but supercilium is more indistinct, two faint buffy wing-bars and underparts softly streaked brownish-grey, greenish gloss on wings.
Similar species Greater Antillean Grackle has longer bill and tail, adults have yellow iris; juvenile has unstreaked underparts and no wing-bars.
Voice Song a series of high warbles and a trill. Call rolling *chuck*.
Habitat and behaviour Secondary woodland, rural gardens and farms. Brood parasite; in the region lays

Adult male has shorter bill and dark iris compared to grackle, May.

in nests of mockingbirds, kingbirds, endemic warblers, vireos and thrushes; the host species raises the cowbird young while host young often starve or are ejected. One pair bred in Thick-billed Vireo nest at West Bay and two young were raised; egg colour said to mimic host species. Feeds on grain, hence present on farms.
Range South America, Panama and Costa Rica. *M. b. minimus* breeds in northern South America and Trinidad and Tobago, and it is established in Florida and the West Indies, where it colonised Puerto Rico in the 1940s, Cuba in the 1980s, Jamaica in 1993, and the Bahamas in 1994.
Status Not a resident or regular breeder up to 2013. Single birds and two breeding records (one successful) on Grand Cayman in 1995; birds reported by North Side farmers from late 1980s. It is expected that attempts to colonise eventually will be successful, as more closed habitats are disturbed, posing a major threat to endemic landbirds.

Adult female, April.

Baltimore Oriole
Icterus galbula

Adult male, February.

Taxonomy Monotypic.
Description L 18–20cm (7–8in). Adult male has black head, back, throat and upper breast; rest of underparts orange, black wings with orange 'shoulders', white wing-bar, bright orange rump and uppertail-coverts, grey bill and legs; in flight shows orange lesser and median wing-coverts, white tips to greater coverts, and orange tail with upper black horizontal band and central stripe. Female has two white-wing-bars; plumage variable and may have blackish feathers on head, nape and upper breast; whitish supercilium and throat, brownish-orange underparts, rump and tail. First-year male has brownish-olive head; brownish-yellow breast, rump and tail; first-year female has olive-grey head and upperparts, yellow breast and tail, grey abdomen and undertail-coverts.
Similar species None.
Voice Usually silent, occasional chattering threat call.
Habitat and behaviour Usually solitary in canopy of trees in woodland and urban/littoral areas.
Range Breeds in North America; winters in south-eastern North America, Middle America to northern South America, and the West Indies: where it is rare in winter, but regular on passage, in the Bahamas, Cuba, Jamaica and Cayman Islands; vagrant to the Lesser Antilles.
Status Rare passage migrant, occasionally uncommon in fall, and rare in winter, September– late April.

Adult female, February.

House Sparrow
Passer domesticus

Adult male, March. Colonised Grand Cayman in 2007.

Taxonomy Polytypic (12). Introduced.

Description L 15cm (5.75in). Non-breeding male has greyish crown, rump and tail; greyish-white cheek patch, black eye-line, black throat, rest of underparts greyish, upperparts rufous and brown with black streaks, pale legs and short, heavy, dark yellowish bill. Breeding male has bright rufous upperparts and nape, black breast, black bill, broad white band on wing. Female has brownish crown with buff supercilium and lores, bill horn, upperparts brown streaked black, and underparts buffy-grey; juvenile similar but duller.

Similar species Female Clay-colored Sparrow has white median crown-stripe, dark malar and lateral stripes on whitish throat. Female Chipping Sparrow has rufous crown and two white wing-bars. Female Dickcissel is larger, with brighter yellow-buff supercilium, larger grey bill and black malar stripe.

Voice *Chirrup* repeated.

Habitat and behaviour Urban areas, foraging for insects, food scraps, seeds. Builds twig nest on ledges of buildings, juveniles in May–June, no other data.

Range Native to Eurasia and Africa. *P. d. domesticus* introduced and breeding in many cities and towns; colonised in the Americas; northern Bahamas, Greater Antilles and Cayman Islands.

Status Introduced new coloniser; small breeding population in George Town, Grand Cayman, from 2007; population has remained static at one site.

Adult female, urban Grand Cayman, June.

APPENDIX 1
Extinct species and subspecies

Grand Cayman Thrush
Turdus ravidus (= Mimocichla ravida)

Local name Old Truss.
Taxonomy Monotypic. The only endemic species in the Cayman Islands.
Description L 24.7cm (9.7in). Plumage uniform bluish-grey to deep grey, lores darker; white abdomen, vent, undertail-coverts and terminal inner web of rectrices; bill, orbital ring, legs and feet bright orange-red.
Range *T. ravidus* was endemic to Grand Cayman, where it was observed during 1886–1938.

Grand Cayman Thrush.

Status Extinct sometime between 1938 and 1965 (Johnston 1969, Bond 1978, PEB). Formerly bred on Grand Cayman only within dry forest on limestone ('cliff'): much of which was felled in the 19th century, probably contributing to its scarcity, and in mangrove forest.

It was reported as common in 1886 by Richardson (Cory 1886). W. W. Brown reported it rare when he took 13 specimens in 1911 (the last collected), and he located it in only two areas of forest which he thought contained the whole population (Bangs 1916). He 'was careful to leave birds to perpetuate the species, if it is not gradually becoming extinct from natural cause, as seems to be the case' (Bangs 1916). The late Bunyon Whittaker remembered taking W. W. Brown to collect specimens in 1911. In 1965 he told D. W. Johnston that there were plenty of 'truss where timber was being cut at the mountain and bulrush walk' (now the Mastic Reserve), and didn't know why they had disappeared. He reported 'the truss was common in cliff… all about…but never saw it after Two-storm' (the 1932 hurricane) (1969 interview by G. B. Reynard, Cornell University). Ira Thompson, a local expert of the same era, never saw the thrush (pers. comm.). **Range** Grand Cayman, Cayman Islands. **Status** *T. ravidus* on Grand Cayman is extinct. It was described as common and collected on 23 April 1886 (Cory 1886); rare, not observed (Nicoll 1904, Lowe 1909, 1911); rare in summer 1911, 13 collected (W. W. Brown, Bangs 1916, Bond 1972). Last field observation was by C. B. Lewis in the present Furtherland Farm area in 1938, during the Oxford Expedition (Johnston 1969). Intensive searches by Johnston in 1965, 1966 and 1967 failed to produce any sightings, and he reported it almost certainly extinct (Johnston 1969). There have been no positive identifications since.

Jamaican Oriole
Icterus leucopteryx

Local name Banana bird.
Taxonomy Polytypic (2). The subspecies in the Cayman Islands was *I. l. bairdi*.
Description L 21cm (8.5in). Adult has bright greenish-yellow upperparts and large white patch on black wing-coverts, black forehead, face, throat and upper breast bib, rest of underparts bright yellow. Pointed bill, tail, legs and feet black.
Range Resident on Jamaica and San Andreas; formerly Grand Cayman, now extirpated.
Status The endemic subspecies *I. l. bardi* is extirpated from Grand Cayman. It was collected on 15 August 1886 (Cory 1886); was 'by no means uncommon' in groups of 5-6 in palms (English 1916); a palm fibre nest with three young at 18m elevation was recorded (Bangs 1916). The last collected specimen was a male at Grape Tree Point on 6 March 1930 (James Bond); observed by C. B. Lewis in 1938 (Johnston *et al.* 1971); the last observation was by the late Bernard St. Aubyn, photographed in his George Town garden in May 1967 (field notes and photograph shown to author). A fossil *Icterus* sp. has been found on Cayman Brac (Morgan 1994).

Jamaican Oriole.

APPENDIX 2
Scientific names of plants mentioned in the text

Family	Common name	Scientific name
Polypodiaceae	Swamp Fern	*Acrostichum aureum*
Ruppiaceae	Ditch grass	*Ruppia* sp.
Cyperaceae	Sedge	*Cyperus* spp., *Eleocharis* spp.
	Sawgrass	*Cladium jamaicense*
Typhaceae	Bulrush/Cattail	*Typha domingensis*
Bromeliaceae	Bromeliads	*Tillandsia* spp.
Palmae	Royal Palm	*Roystonea regia*
	Silver Thatch	*Coccothrinax proctorii*
	Coconut Palm	*Cocos nucifera*
Agavaceae	Corato, Century Plant, Agave	*Agave caymanensis*
Smilacaceae	Wire Wiss, Greenbrier	*Smilax havanensis*
Orchidaceae	Banana Orchid	*Myrmecophila thomsoniana*
Canellaceae	Pepper Cinnamon	*Canella winterana*
Moraceae	Wild Fig	*Ficus citrifolia*
	Wild Fig	*Ficus aurea*
Casuarinaceae	Casuarina or Australian Pine	*Casuarina equisetifolia*
Nyctaginaceae	Cabbage Tree	*Guapira discolor*
Cactaceae	Royen's Tree Cactus	*Pilosocereus swartzii*
	Vine Pear	*Selenicereus grandiflorus*
Aizoaceae	Sea-purslane	*Sesuvium portulacastrum*
Chenopodiaceae	Dwarf Glasswort	*Salicornia bigelovii*
Bataceae	Saltwort, Turtleweed	*Batis maritima*
Polygonaceae	Sea-grape	*Coccoloba uvifera*
Clusiaceae	Balsam	*Clusia rosea*
Malvaceae	Plopnut	*Thespesia populnea*
	Cotton	*Gossypium* sp.
Salicaceae	Shake Hand	*Xylosma bahamense*
Caricaceae	Pawpaw, Papaya	*Carica papaya*
Sapotaceae	Mastic	*Sideroxylon foetidissimum*
	Wild Sapodilla	*Sideroxylon salicifolium*
	Shake Hand, White Thorn	*Sideroxylon horridum*
	Neesberry	*Manilkara zapota*
Leguminosae	Sea Bean	*Canavalia rosea*

Family	Common name	Scientific name
	Logwood	*Haematoxylum campechianum*
	Tamarind	*Tamarindus indica*
Myrtaceae	Bastard Strawberry	*Calyptranthes pallens*
	Cherry	*Myrcianthes fragrans*
	Strawberry	*Eugenia axillaris*
Combretaceae	Almond	*Terminalia catappa*
	Black Mastic	*Terminalia eriostachya*
	Buttonwood	*Conocarpus erectus*
	White Mangrove	*Laguncularia racemosa*
Rhizophoraceae	Red Mangrove	*Rhizophora mangle*
Viscaceae	Scorn-the-ground	*Phoradenrdon rubrum*
Euphorbiaceae	Wild Coco-plum	*Savia erythroxyloides*
	Bitter Plum	*Picrodendron baccatum*
	Narrow-leaf Ironwood, Crab Bush	*Gymnanthes lucida*
	Manchineel	*Hippomane mancinella*
Sapindaceae	Pompero	*Hypelate trifoliata*
	Wild Ginep	*Exothea paniculata*
Burseraceae	Red Birch	*Bursera simaruba*
Anacardiaceae	Mango	*Mangifera indica*
	Maiden Plum	*Comocladia dentata*
Rutaceae	Candlewood	*Amyris elemifera*
Meliaceae	Bastard Mahogany	*Trichilia glabra*
	Mahogany	*Swietenia mahagoni*
	Cedar	*Cedrela odorata*
Erythroxylaceae	Smoke Wood	*Erythroxylum areolatum*
Apocynaceae	Wild Jasmine	*Plumeria obtusa*
Convolvulaceae	Bay Vine	*Ipomoea pes-caprae*
Menyanthaceae	Water Snowflake	*Nymphoides indica*
Boraginaceae	Broadleaf	*Cordia sebestena var. caymanensis*
	Spanish Elm	*Cordia gerascanthus*
Verbenaceae	Fiddlewood	*Petitia domingensis*
Avicenniaceae	Black Mangrove	*Avicennia germinans*
Oleaceae	Ironwood	*Chionanthus caymanensis*
Scrophulariaceae	[a false foxglove]	*Agalinis kingsii*
Bignoniaceae	Whitewood	*Tabebuia heterophylla*
Rubiaceae	Bastard Ironwood	*Exostema caribaeum*
Asteraceae	[Siam Weed]	*Chromolaena odorata*

APPENDIX 3
Distribution of breeding seabirds, waterbirds and shorebirds in the Cayman Islands

	Grand Cayman		Cayman Brac		Little Cayman	
	RB	MB	RB	MB	RB	MB
Pied-billed Grebe	+		+		+	
White-tailed Tropicbird		+		+		
Brown Booby			+			
Red-footed Booby					+	
Magnificent Frigatebird					+	
Least Bittern	+					
Snowy Egret	+				+	
Little Blue Heron	IB				IB	
Tricolored Heron	+		+		+	
Cattle Egret	+		+		+	
Green Heron	+		+		+	
Yellow-crowned Night-Heron	+		+		+	
West Indian Whistling-Duck	+		+		+	
Purple Gallinule	+		IB		IB	
Common Moorhen	+		+			
American Coot	+		+			
Black-necked Stilt	+		+			
Willet	+		+			
Least Tern		+		FB		FB
Bridled Tern		+				
Totals	14		12		15	

RB   Resident breeder
MB   Migrant breeder (summer)
IB   Intermittent breeding; majority are North American migrants in winter
FB   Former breeder; no recent records

APPENDIX 4
Distribution of breeding landbirds in the Cayman Islands

	Grand Cayman		Cayman Brac		Little Cayman	
	RB	MB	RB	MB	RB	MB
White-crowned Pigeon	+		+		+	
White-winged Dove	+		+		+	
Zenaida Dove	+		+		+	
Common Ground Dove	+		+		+	
Caribbean Dove	+					
Cuban Parrot	+		+			
Mangrove Cuckoo	+		+		+	
Smooth-billed Ani	+		+		+	
Barn Owl	+		+		+	
Short-eared Owl	+		+		?B	
Antillean Nighthawk		+		+		+
West Indian Woodpecker	+					
Northern Flicker	+					
Caribbean Elaenia	+		+		+	
La Sagra's Flycatcher	+					
Gray Kingbird		+		+		+
Loggerhead Kingbird	+		+			
Red-legged Thrush			+			
Northern Mockingbird	+		+		+	
Thick-billed Vireo	+		+			
Black-whiskered Vireo				+		+
Yucatan Vireo	+					
Yellow Warbler	+		+		+	
Vitelline Warbler	+		+		+	
Bananaquit	+		+		+	

	Grand Cayman		Cayman Brac		Little Cayman	
	RB	MB	RB	MB	RB	MB
Western Spindalis	+					
Cuban Bullfinch	+					
Yellow-faced Grassquit	+		+		+	
Greater Antillean Grackle	+				+	
Totals	**27**		**20**		**17**	

RB   Resident breeding
MB   Migrant breeding (summer)
?B   Breeding not confirmed

Cuban Bullfinch.

APPENDIX 5
Distribution of the 17 endemic subspecies of landbirds found on Grand Cayman (GC), Little Cayman (LC) and Cayman Brac (CB)

Landbirds		Endemic subspecies
Caribbean Dove	*Leptotila jamaicensis*	*collaris* (GC)
Cuban Parrot	*Amazona leucocephala*	*caymanensis* (GC) *hesterna* (CB)
West Indian Woodpecker	*Melanerpes superciliaris*	*caymanensis* (GC)
Northern Flicker	*Colaptes auratus*	*gundlachi* (GC)
Caribbean Elaenia	*Elaenia martinica*	*caymanensis* (GC, LC, CB)
Loggerhead Kingbird	*Tyrannus caudifasciatus*	*caymanensis* (GC, CB)
Red-legged Thrush	*Turdus plumbeus*	*coryi* (CB)
Thick-billed Vireo	*Vireo crassirostris*	*alleni* (GC, CB)
Yucatan Vireo	*Vireo magister*	*caymanensis* (GC)
Bananaquit	*Coereba flaveola*	*sharpei* (GC, LC, CB)
Vitelline Warbler	*Dendroica vitellina*	*vitellina* (GC) *crawfordi* (LC, CB)
Western Spindalis	*Spindalis zena*	*salvini* (GC)
Cuban Bullfinch	*Melopyrrha nigra*	*taylori* (GC)
Greater Antillean Grackle	*Quiscalus niger*	*caymanensis* (GC) *bangsi* (LC)

APPENDIX 6
Vagrants and very rare migrants to the Cayman Islands

Species with ten or fewer local records

Canada Goose   *Branta canadensis*
Taxonomy Polytypic (11).
Description L 64–110cm (25–43in). Head and long neck black with wide white chin 'strap'; back and wings banded brownish-grey, pale breast, brownish-grey underparts darker on flanks, white rump and undertail-coverts, dark subterminal band on white tail.
Range Breeds in Canada and Alaska; natural populations migrate south in winter to the lower United States, Mexico and in the West Indies in northern Bahamas: three records in Cuba, one in Hispaniola, no modern records in Jamaica. Feral populations are sedentary.
Status Vagrant. One record of an adult on Cayman Brac, May 2011.

Wood Duck   *Aix sponsa*
Taxonomy Monotypic.
Description L 45–51cm (17–20in). Male has unmistakable green crest, white 'straps' on face and neck, orange and black bill, rufous-plum breast and bright yellow-buff sides and flanks. Female and eclipse male are crested, small grey bill, wide white asymmetrical eye-ring, underparts brown spotted.
Similar species Female Ring-necked Duck is dark brown overall, white eye-ring is much smaller, blue bill is wider with white ring and black tip.
Range Breeds in North America south to the Gulf coast and southern Florida, and the West Indies in Cuba; winters in North America, Cuba and, rarely, in the Bahamas.
Status Vagrant. One record of eight females on Little Cayman, February-March 1984. No recent records; familiar to local hunters in the 1960s.

Mallard   *Anas platyrhynchos*
Taxonomy Polytypic (7).
Description L 51–71cm (10–28in). Female and eclipse male are mottled brown and yellowish-buff; female has orange bill with black markings, eclipse male and juvenile have blackish bill. Male has iridescent green head and neck, white collar, chestnut breast, yellow bill and legs. Shows violet-blue speculum with white border in flight.
Similar species Male Northern Shoveler has long spatulate bill, white breast and chestnut sides, shows green speculum and blue forewing in flight. Female Gadwall has grey bill with yellow sides and white speculum in flight.
Range Four subspecies breed in North America; winters in North America south to the Gulf coast, Mexico and rarely in the West Indies: in the Bahamas and western Cuba. Introduced populations in the Bahamas and Antilles. Holarctic breeder.
Status Vagrant. No recent records. Now domesticated on Grand Cayman.

White-cheeked Pintail   *Anas bahamensis*
Taxonomy Polytypic (3).
Description L 38–48cm (15–19in). Elegant slender duck with long buffy tail, light and dark brown plumage, diagnostic white cheeks and throat, red at base of grey bill. Shows green speculum with buffy borders in flight.
Range *A. b. bahamensis* breeds in South America and the West Indies in the Bahamas, Greater Antilles and some islands in the Lesser Antilles; vagrant to Jamaica. Two subspecies breed in South America.
Status Vagrant. Two records in winter on Grand Cayman: 1985, 1986; no recent records.

Redhead   *Aythya americana*
Taxonomy Monotypic.
Description L 46–56cm (18–22in). Medium-sized diving duck with steeply rising forehead and rounded head. Eclipse male and female have blue bill with white subterminal band and black tip, plumage is brown with lighter brown bands on flanks. Female has white eye-ring. Black neck and breast develop in early winter on eclipse and juvenile males. Breeding male has bright reddish head, pale grey back; black neck, breast, rump and tail-coverts.
Range Breeds in North America; winters south to Guatemala and the West Indies in the Bahamas and Cuba, vagrant to Jamaica.
Status Vagrant. One record of two males and 13 females that overwintered at Red Bay, Grand Cayman, December 2011–February 2012.

Hooded Merganser   *Lophodytes cucullatus*
Taxonomy Monotypic.
Description L 40–48cm (16–19in). Slender, with long body and tail, anvil-shaped head and thin, dark, serrated hooked bill. Male is unmistakable with high black crest with large central white patch when raised, white breast and orange-cinnamon sides. Female, eclipse male and immature have rounded bushy greyish-brown crest, dark brown back, greyish underparts and dark bill orange at base.
Similar species Female Red-breasted Merganser has grey upperparts, whitish throat and breast, light orange crest, bill is longer and red; shows white inner secondaries in flight.
Range Breeds in North America; winters in the United States including Florida and the Gulf coast to eastern Mexico and the West Indies: rare in northern Bahamas and Cuba, and vagrant elsewhere.
Status Vagrant. One record on Grand Cayman, January 1996; no recent records.

Masked Duck   *Nomonyx dominicus*
Taxonomy Monotypic.
Description L 30–36cm (12–14in). Small stiff-tailed duck with flat crown and diagnostic white patch on upperwing in flight. Female, eclipse male and juvenile have plumage barred brownish-grey, black crown and two black stripes across buffy face, bill grey. Breeding male has black face, cinnamon-rufous plumage with black spotting on flanks and black tip to blue bill.
Similar species Ruddy Duck shows dark wings in flight. Male has black head with white cheeks and entirely blue bill. Female and immature have single dark line across cheeks.
Range Resident locally in North America on the United States Gulf coast, Middle America to South America, and the West Indies in the Bahamas (rare), Greater Antilles (locally common in Cuba and Jamaica, rare in Puerto Rico), and Lesser Antilles. A nomadic and irregular wanderer (AOU 1998).
Status Vagrant. One record on Grand Cayman, 1972 by Wetmore; not recently reported. A former breeder (Savage English 1916).

Least Grebe   *Tachybaptus dominicus*
Taxonomy Polytypic (4).
Description L 23–26cm (9–10in). Small, blackish grebe with dark pointed bill, yellow-orange iris, white undertail-coverts; shows white wing patch in flight.
Similar Species Pied-billed Grebe is larger with brownish-grey plumage and black band around large pale bill.
Range *T. d dominicus* breeds from coastal southern Texas through Middle America to South America and the West Indies: in the Bahamas (except Grand Bahama) and Greater Antilles, where common in Cuba and Jamaica, uncommon Hispaniola and Puerto Rico; mainly sedentary.
Status Vagrant. One record on Grand Cayman, winter 1995–1996; no recent records.

Black-capped Petrel   *Pterodroma hasitata*
Taxonomy Polytypic (2).
Description 35–40cm (14–16in). Black crown extends to beyond eye and to upper breast forming a partial collar, brownish-grey upperparts and tail except for white forehead, hindneck, rump and uppertail-coverts; white underparts. In flight blackish leading edge and narrow trailing edge to white

underwing, wing bent at 'wrist'.

Similar species Great Shearwater has white underwing and black forehead.

Range Breeds in the West Indies in Hispaniola, and formerly in Guadeloupe and Dominica. Occurs widely at sea in West Indies and eastern North American waters away from the breeding grounds.

Status Pelagic casual visitor. One record by James Bond at sea off Cayman Brac, April 1961; most likely under-reported from Cayman waters.

Great Shearwater   *Puffinus gravis*

Taxonomy Monotypic.

Description 48cm (19in). The largest shearwater. Greyish-brown head, upperparts and tail, some white on hindneck, white rump and whitish underparts.

Similar species Black-capped Petrel is smaller with white forehead and blackish leading edge to underwing.

Range Casual visitor in summer to the Gulf of Mexico and offshore on the eastern United States, the Bahamas and eastern Antilles. Breeds on islands in Southern Ocean.

Status Vagrant. Two records of dead birds on Grand Cayman: 1992, 2002.

Audubon's Shearwater   *Puffinus lherminieri*

Taxonomy Polytypic (10).

Description 30cm (12in). Small shearwater with entirely blackish-brown upperparts, white underparts except for blackish-edged underwings and blackish undertail-coverts.

Range Breeds locally in the Caribbean on cays and islands off Panama, Nicaragua and Venezuela and in the West Indies in the Bahamas (common), Puerto Rico, Virgin Islands and Lesser Antilles. Pantropical.

Status Casual visitor. Two records at sea, and one photographed in North Sound, Grand Cayman in January 2002. Former breeder on Cayman Brac: extirpated before permanent settlement.

American White Pelican   *Pelecanus erythrorhynchos*

Taxonomy Monotypic.

Description L 150cm (60in). Very large, heavy white bird with huge yellow-pink bill and orange legs; in flight shows black primaries and secondaries.

Range Breeds in North America and winters in southern United States to Costa Rica and rarely, in the West Indies in Cuba and Puerto Rico; otherwise vagrant.

Status Vagrant. One 1977 record in North Sound, Grand Cayman by James Bond; not recently reported.

American Bittern   *Botaurus lentiginosus*

Taxonomy Monotypic.

Description L 58–62cm (23–24in). Medium-sized heron. Face and throat cream and buff, black malar stripe continues to sides of breast, plumage variable rich, soft browns with darkly streaked underparts, bill and legs yellowish-brown. In flight shows blackish primaries.

Similar species Immature night-herons have less streaking on underparts and shorter, rounded wings with pale spots.

Range Breeds from North America to central Mexico; winters from North America to southern Mexico and the West Indies in Swan Islands, Cuba and the Bahamas, rare in Hispaniola, Puerto Rico, and vagrant to the Cayman Islands, Jamaica and Lesser Antilles.

Status Vagrant. Three records of single birds recorded on grass airstrips in October, January-March 1984, 1997; no recent reports.

American Golden Plover   *Pluvialis dominica*

Taxonomy Monotypic.

Description L 26cm (10in). Large plover. Non-breeding adult has greyish and gold upperparts, dark crown, whitish supercilium, white underparts with streaking on breast, black bill and legs. Breeding adult has mottled gold, black and brown upperparts, white S-shaped stripe from forehead and crown to side of breast, entirely black underparts. In flight shows dark rump, tail and wings, and unmarked axillaries.

Similar species Black-bellied Plover is larger with heavier bill, white rump and wing-stripe, and black axillaries.

Range Breeds in northern North America; winters in southern South America. Rare passage migrant in the West Indies.

Status Rare passage migrant found on herbaceous wetlands and flooded grassland. Eight records: in October and March-April, 1984–2009.

Snowy Plover   *Charadrius nivosus*

Taxonomy Polytypic (5).

Description L 15 cm (5.75in). Small plover, with pale sandy-grey upperparts, short wings, long grey legs, thin black bill, and white forehead, supercilium, collar and underparts. Breeding male has black bar across forecrown, ear-coverts patch and bar on sides of neck; these dark brown in female. In winter plumage black areas become grey-brown.

Similar species Wilson's and Semipalmated Plovers have brown upperparts and complete breast bands.

Range Breeds in the United States through Middle America to Panama. *C. n. nivosus* breeds in the West Indies in the southern Bahamas, Turks and Caicos Islands, Cuba, Hispaniola, locally in Puerto Rico, Virgin Islands and Lesser Antilles. Winters on the Gulf coast to Middle America (casually to Panama), the Bahamas and Greater Antilles, except Jamaica and the Cayman Islands, where it is a vagrant.

Status Vagrant. One record on Grand Cayman, November 1991 to January 1992; no recent records.

Marbled Godwit   *Limosa fedoa*

Taxonomy Polytypic (2).

Description L 40–51cm (16–20in). Large shorebird with very long, slightly upturned, bicoloured bill (pink with dark tip) and long dark legs. Non-breeding has upperparts spotted and barred cinnamon, dark brown and black, and underparts plain buff. Breeding adult has brighter upperparts and streaking on breast. Shows cinnamon underwings in flight.

Range *L. f. fedoa* breeds in the northern North America; winters to southern United States coasts through Middle America to north-western South America. Rare in the West Indies, mainly in fall passage.

Status Rare passage migrant. Two records on Grand Cayman, in September 1992, 2002.

Baird's Sandpiper   *Calidris bairdii*

Taxonomy Monotypic.

Description L 19cm (7.5in). Non-breeding adult has grey-brown upperparts with pale edgings, buffy breast finely streaked, rest of underparts white, short legs. Breeding adult has upperparts silvery cinnamon and buff with black centres to feathers, bright buff on face, buffy breast streaked and sharply demarcated from white abdomen, short straight black bill, wings extending well beyond tail at rest. In flight shows dark rump and tail with white sides.

Similar species White-rumped Sandpiper has white rump.

Range Breeds in North America north of 60°N (also in Greenland and Siberia); winters in southern South America. Very rare on passage in the West Indies; only recorded from Cayman Islands and Virgin Islands, Dominica, St Lucia, the Grenadines and Barbados.

Status Rare passage migrant. Six records on Grand Cayman and one on Cayman Brac, in September, October and May, 1992–2012. Birds have appeared in damp grassland, herbaceous wetlands and disturbed mangrove.

Ruff   *Philomachus pugnax*

Taxonomy Monotypic.

Description L 28 cm (11in). Small-headed wader with scaly markings on upperparts; shows oval white sides to rump in flight, legs yellow or orange. Underparts greyish or whitish in non-breeding adults, buffish in juveniles. Males much larger than females; only breeding males show unmistakable crest and ruff; breast blotched black in breeding female.

Range Holarctic. Breeds in Arctic and subarctic Eurasia (plus a nesting record in Alaska); winters Europe, Middle East, southern Africa, and across southern Asia to Australia. Regular in migration on east coast

North America to Florida, and casual throughout, also the Bahamas, Greater and Lesser Antilles.
Status Vagrant. First record of male in partial breeding plumage, flooded grassland at the airport, Cayman Brac, following an extensive regional tropical disturbance, 22–23 May 2012.

Buff-breasted Sandpiper *Tryngites subruficollis*
Taxonomy Monotypic.
Description L 19–22cm (7.5–8.5in). Resembles a plover. Slender with small round head, dark crown and dark eye; face and underparts warm buff with spots on sides of neck and breast; upperparts with dark brown centres and pale edgings to feathers, black pointed bill and yellow legs.
Range Breeds in arctic North America north of 60° N; winters in southern South America. Rare on passage, mainly in fall, in the West Indies in the Lesser Antilles and Cayman Islands, casual or vagrant to the Bahamas and Greater Antilles.
Status Rare passage migrant. Two records on Cayman Brac, September 1984, May 2002, and two on Grand Cayman in fall and one in May 2012. Birds have been found on on grass edges of airport runways.

Bonaparte's Gull *Larus philadelphia*
Taxonomy Monotypic.
Description L 30–35cm (12–14in). Small tern-like gull with thin, straight black bill. Non-breeding adult and first-winter are white with diagnostic black spot on ear coverts. Non-breeding adult has pale grey mantle, white tail, white outer primaries on upperwing, and pink legs; in flight white primaries edged with black on trailing edge of wing. Breeding adult has black head and orange-red legs. In flight first winter shows dark upperwing-covert bar, dark tips to primaries and secondaries, narrow black terminal tail-band.
Similar species Laughing Gull is larger with dark grey mantle, black wing tips, reddish bill and black tail; first-year has dark brown and blackish mantle and broad terminal tail-band.
Range Breeds in northern North America. Winters on Pacific, Atlantic and Gulf coasts south to Mexico and uncommonly to the Bahamas and Cuba.
Status Vagrant. One photographed on Little Cayman, December 2010.

Lesser Black-backed Gull *Larus fuscus*
Taxonomy Polytypic (4).
Description L 53–63cm (21–25in). Large slender gull. Non-breeding adult has white head and hindneck finely streaked brown, grey-black mantle, yellow bill with red spot near tip of lower mandible, white rump and tail, pale yellow legs and feet. First-winter patterned grey-brown above, similar to Herring Gull but greyer; head and underparts paler, rump and uppertail-coverts whitish, barred darker; black bill, dark tail band. Second winter has whiter head, underparts and uppertail coverts.
Similar species Herring Gull is larger; juvenile is darker but adult is paler, with pink legs.
Range Breeds in Palearctic. *L. f. graellsii* is a regular visitor to eastern North America, and the northern Bahamas.
Status Vagrant. One record on Little Cayman, February 2009.

Brown Noddy *Anous stolidus*
Taxonomy Polytypic (5).
Description L 38–40cm (15–16in). Adult is entirely blackish-brown, except for silvery-white crown shading to grey hind neck, white crescent below eye, tail 'double-rounded' due to shorter inner tail feathers, long bill. Immature is dark brown with lighter brown coverts, no white on head.
Similar species Black Noddy has not been recorded.
Range *S. s. solidus* breeds in the Gulf-Caribbean cays throughout the West Indies; winters at sea in the vicinity of the breeding grounds and wanders widely after storms; it also breeds in South Atlantic islands and Gulf of Guinea to Cameroon. Another subspecies breeds on islands off the Pacific coast of Mexico and Middle America. Pantropical.
Status Vagrant or rare passage migrant. Six records of single birds, Grand Cayman and Little Cayman, July-October, 1995-1996, and 90-100 on Vidal Cay, Grand Cayman, late June –5 July 2012.

Long-tailed Jaeger   *Stercorarius longicaudus*
Alternative name Long-tailed Skua.
Taxonomy Polytypic (2).
Description L 50–58cm (19.5–23in). Slender, with long pointed wings, long tail and short thick bill. Adult has rounded head with blackish cap, pale face, neck and underparts with barring on flanks and undertail-coverts, brownish-grey upperparts and long streaming central tail feathers; in flight shows dark underwings with contrasting white bases to outer primaries. Immature lacks cap and elongated tail feathers; plumage barred.
Range In Nearctic breeds from Alaska to northern Quebec; winters in the Pacific and Atlantic oceans south of 40°N. *S. l. pallescens* is mainly rare passage migrant to the West Indies, although a few overwinter in Lesser Antillean waters.
Status Vagrant. One record from Cayman Islands' waters by James Bond in 1961.

Black-billed Cuckoo   *Coccyzus erythropthalmus*
Taxonomy Monotypic.
Description L 30cm (12in). Slender with grey-brown upperparts, red eye-ring, and thin, decurved blackish bill; whitish underparts and small spots on underside of greyish tail.
Similar species Yellow-billed Cuckoo has thicker bill with lower mandible yellow, large white tail spots and bright rufous primaries. Mangrove Cuckoo has black ear patch and buffy-cinnamon underparts.
Range Breeds in North America; winters in north-western South America; rare on passage in the West Indies in the northern Bahamas, Cuba and Hispaniola, vagrant elsewhere.
Status Vagrant or rare passage migrant. Two records on Grand Cayman, September 2002 and October 1993.

Common Nighthawk   *Chordeiles minor*
Taxonomy Polytypic (9).
Description L 22–25cm (9–10in). Smaller than Antillean Nighthawk and wings do not extend beyond the tail at rest; plumage nearly identical. Upperparts blackish-brown speckled with whitish-grey, cinnamon, russet and black; underparts barred whitish grey; large flat head, whitish supercilium, small bill with a wide gape and small weak legs. In flight, long, slender pointed blackish wings bent in a V at the carpal joint and blackish primaries. Male has white throat patch, broad white cross-band on five outer primaries, and broad white subterminal band on tail. Female throat patch smaller and buffy, less distinct white band on primaries, underparts buffy rather than greyish, white subterminal band absent. Immature similar to female but underparts heavily barred.
Similar species Antillean Nighthawk is slightly larger but otherwise similar; only separable by call.
Voice Two syllable call *pee-nt*, seldom heard on migration.
Range Breeds from North to Middle America; winters in South America. Most migrate through Middle America and uncommon in passage in the Greater Antilles and Cayman Islands.
Status Rare passage migrant in October–November and April–May. Late fall birds may be identified by call after breeding Antillean Nighthawk have migrated. Occurs in same habitat as Antillean Nighthawk.

Acadian Flycatcher   *Empidonax virescens*
Taxonomy Monotypic.
Description L 13cm (5.25in). Flat forehead with peaked crown, greenish upperparts, yellowish underparts with olive wash on sides of breast, whitish throat; narrow, pale yellow eye-ring, broad bill with yellow lower mandible, greyish legs, two sharply defined whitish wing-bars on long wings. Immature has yellowish wash on underparts.
Similar species Least Flycatcher is smaller with shorter wings, conspicuous white eye-ring and dark bill.
Range Breeds in the eastern United States; winters in Middle America and north-western South America; rare on passage in the northern Bahamas, western Cuba and Cayman Islands in fall.
Status Vagrant or rare passage migrant on Grand Cayman: three records in fall, 1994, 2002, 2008.

Least Flycatcher   *Empidonax minimus*

Taxonomy Monotypic.
Description L 13cm (5.25in). Large head, greyish upperparts, short bill, short wings with two sharply defined white wing-bars, conspicuous white eye-ring, olive wash on breast; rest of underparts pale.
Similar species Acadian Flycatcher is larger, with longer wings, greenish upperparts, yellow lower mandible and pale yellow narrow eye-ring.
Range Breeds in North America; winters in Middle America to Panama; vagrant to the West Indies.
Status Vagrant. Three records on Grand Cayman, February and March 1971; no recent records. One specimen collected 1904.

Eastern Phoebe   *Sayornis phoebe*

Taxonomy Monotypic.
Description L 18cm (7in). Greyish-brown upperparts, long dark tail, no eye-ring or wing-bars, black bill, dark breast sides, whitish underparts with pale yellowish wash on lower abdomen.
Range Breeds in North America; winters in North America including Florida, and Mexico; casual in winter in the northern Bahamas and Cuba.
Status Vagrant or casual visitor. One record on Little Cayman, November 1988; no recent records.

Great Crested Flycatcher   *Myiarchus crinitus*

Taxonomy Monotypic.
Description L 20cm (8in). Large myiarchian flycatcher with defined crest, olive-brown head and back, grey throat and breast, abdomen bright yellow, bill black with orange inner lower mandible, whitish wing-bars with rufous outer webs to primaries; extensively rufous tail.
Similar species La Sagra's Flycatcher has greyish-white underparts.
Range Breeds in North America; winters in Florida and Middle America to northern South America. Very rare on passage in Cuba.
Status Vagrant or very rare passage migrant. Two records on Grand Cayman, September 2002 and October 2009, following tropical storms.

Tropical Kingbird   *Tyrannus melancholicus*

Taxonomy Polytypic (3).
Description L 23cm (9.25in). Large tyrannid with greenish back, buffy edges to wing-coverts, pale grey head with blackish mask through eye, white throat, pale grey breast, yellow abdomen and undertail-coverts, long black bill, notched brownish-black tail.
Similar species Great Crested Flycatcher has dark upperparts, rufous on wings and tail.
Range Breeds in Arizona in North America, Middle America and South America; winters in the breeding range; *T. m. satrapa* is casual in Cuba.
Status Vagrant. One record on Grand Cayman, October 1995, following a tropical storm.

Blue-headed Vireo   *Vireo solitarius*

Taxonomy Polytypic (2).
Description L 14cm (5.5in). Olive-green upperparts; bluish-grey crown, nape and ear-coverts; white eye-ring and lores form 'spectacles', two bright white wing-bars; white below with yellowish flanks.
Similar species White-eyed Vireo has white iris and yellow spectacles. Thick-billed Vireo has throat and underparts yellowish-buff and yellow lores. Yellow-throated Vireo has yellow eye-ring, throat and breast.
Range Breeds in North America; *V. s. solitarius* winters in the southern United States, Middle America and, rarely, in Cuba.
Status Vagrant. One record on Little Cayman, October 2002.

Philadelphia Vireo   *Vireo philadelphicus*

Taxonomy Monotypic.
Description L 13cm (5.25in). Olive-grey upperparts and dark grey crown, broad white supercilium,

Philadelphia Vireo.

dark eye-line to lores, brown iris, throat yellow and rest of underparts whitish. Unmarked wings with no wing-bars are diagnostic.

Similar species Red-eyed Vireo has dark edges to white supercilium and red iris.

Range Breeds in northern North America; winters from Middle America to Panama; rare passage migrant in the Bahamas, Cuba, Jamaica and Cayman Islands.

Status Rare passage migrant. Three records on Grand Cayman in October, late April and May, 2002–2009.

European Starling   *Sturnus vulgaris*

Taxonomy Polytypic (12).

Description L 15cm (5.75in). Non-breeding adult is heavily spotted with white. Breeding adult glossy black with purple iridescence on head, neck and coverts; white spots on mantle, bright yellow bill and pink legs. Juvenile is brownish-grey with dark bill.

Range Old World species. Introduced and breeding in North America, the Bahamas, Jamaica, Puerto Rico and the Virgin Islands. North American birds winter in the northern Bahamas; vagrant to Cuba.

Status Vagrant, and potential coloniser. Four records on Cayman Brac and Little Cayman, March and April, 1987–1998. No recent records.

Golden-winged Warbler   *Vermivora chrysoptera*

Taxonomy Monotypic.

Description L 13cm (5.25in). Blue-grey upperparts with yellow crown and wing-coverts, greyish underparts except for white undertail-coverts. Male has wide white supercilium; black throat and ear-coverts are separated by broad white stripe. Female is paler with face and throat markings dark grey, crown greenish-yellow.

Range Breeds in North America; winters from Mexico to northern South America and rarely in the West Indies: rare on passage in Cuba, Hispaniola and Cayman Islands; vagrant to Jamaica.

Status Rare passage migrant. Over ten records on Grand Cayman, September-November and one record in March, 1991–2002. Has appeared in littoral woodland, dry and mangrove shrubland, and gardens.

Orange-crowned Warbler   *Oreothlypis celata*
Taxonomy Polytypic (4).

Description L 13cm (5.25in). Olive-grey upperparts with concealed orange crown patch, yellowish-green supercilium, short dark eye-line and yellowish broken eye-ring, sharply pointed bill, no wing-bars, greenish buff underparts with faint streaking, and yellow undertail-coverts.

Similar species No other warbler has a combination of grey head, yellow undertail-coverts and no wing-bars.

Range Breeds widely in North America; winters in the south-eastern United States to Mexico and, rarely, in the Bahamas; vagrant to Cuba and Jamaica.

Status Vagrant. Three records on Grand Cayman in October, 1993–1995; no recent records.

Nashville Warbler   *Oreothlypis ruficapilla*
Taxonomy Polytypic (2).

Description L 12cm (4.75in). Bluish-grey head and neck, olive-green back and wing-coverts, pronounced white eye-ring, no wing-bars, underparts yellow except for small white area on abdomen.

Range *O. r. ruficapilla* breeds in eastern North America; winters mainly in Mexico, also in the West Indies: where it is rare in winter and on passage in the Bahamas; vagrant to Cuba and Jamaica. *O. r ridgewayi* breeds and winters in western North America.

Status Rare passage migrant. Nine records, September–November, February and April, 1992–2009. Has appeared in littoral woodland and mangrove shrubland.

Connecticut Warbler   *Oporornis agilis*
Taxonomy Monotypic.

Description L 14cm (5.5in). Both adults have white eye-ring, no wing-bars, pale legs and short tail. Male has grey head and upper breast forming a hood, brownish upperparts, pale lower mandible, rest of underparts yellow. Female is duller with brownish hood.

Range Breeds in northern North America; winters in north-central South America. On passage in both directions through the West Indies where it is very rare in the Bahamas, western Greater Antilles and Cayman Islands.

Status Rare passage migrant. Four records on Grand Cayman, September, 1992–2008. Thrush-like, secretive and terrestrial in dry and buttonwood shrubland.

Mourning Warbler   *Geothlypis philadelphia*
Taxonomy Monotypic.

Description L 14cm (5.5in). Male has dark blue-grey head, neck and throat; black patch on grey breast, dark eye and yellow underparts; no wing-bars. Female is paler with greyish throat patch and partial white eye-ring. Immature has whitish eye-ring and yellowish throat and patch on abdomen. All have pale legs.

Range Breeds in North America; winters southern Middle America to northern South America. Casual in the Bahamas, and vagrant to Cuba and Jamaica; occasionally overwinters in Hispaniola.

Status Vagrant. One record on Grand Cayman, September 2009.

Cerulean Warbler   *Setophaga cerulea*
Taxonomy Monotypic.

Description L 12cm (4.5in). Adults have two white wing-bars, pointed wings and short tail. Male has bright blue upperparts, white underparts with dark collar and streaked sides. Female is blue-green above with whitish-buff supercilium, underparts with faint streaking; immature is similar with yellowish supercilium and underparts.

Similar species Black-throated Blue Warbler male has a black face.
Range Breeds in North America; winters in north-western South America. Most migrate trans-Gulf through the Yucatan and Panama and low numbers pass through the West Indies: where it is rare in the Bahamas, western Cuba and Cayman Islands; vagrant to Jamaica.
Status Rare passage migrant. There are six records on Grand Cayman and Cayman Brac, August-October, and April; one in December is the only winter record for the region. Global Conservation status: Vulnerable. Has appeared in littoral/urban areas, dry forest and shrubland.

Pine Warbler   *Setophaga pinus*
Taxonomy Polytypic (4).
Description L 14cm (5.5in). Male has olive-green upperparts, diagnostic pale neck patch, dark head and ear-coverts, broken yellow eye-ring and two white wing-bars; yellow throat and breast, white abdomen and undertail-coverts, olive streaking on sides of breast and flanks. Female and immature have dull brown upperparts and greyer underparts.
Similar species Yellow-throated and White-eyed Vireos have bright yellow 'spectacles', lack streaking on sides and dark ear patches. The latter has a white throat and breast.
Range *S. p. pinus* is migratory, breeding in south-eastern North America and wintering in the southern breeding range, rarely south to the Bahamas and Greater Antilles; vagrant to Cuba and Jamaica. Three sedentary resident subspecies occur in Florida, the northern Bahamas and Hispaniola.
Status Vagrant. Five records on Grand Cayman in October, 1982–2003. Has appeared in Casuarina Pines in littoral/urban woodland and in dry shrubland.

Wilson's Warbler   *Cardellina pusilla*
Taxonomy Polytypic (3).
Description L 12cm (4.5in). Greenish-olive upperparts, black iris with white eye-ring in yellow face, yellow underparts, short bill and pale legs. Male has black crown cap; female has olive crown and yellow supercilium.
Similar species Female Yellow Warbler is larger with yellow tail spots and no supercilium.
Range Breeds in North America; winters from the Gulf coast through Middle America to Panama. Casual on passage and in winter in the Bahamas and Greater Antilles where rare on Jamaica, Cuba and Cayman Islands.
Status Rare passage migrant. Five records, September-October, 1995–2002. Has appeared in littoral/urban woodland and in dry shrubland.

Yellow-breasted Chat   *Icteria virens*
Taxonomy Polytypic (2).
Description L 19cm (7.5in). Large warbler with thick black bill and long tail, olive upperparts; white eye-ring, supercilium and malar stripe, black lores, orange-yellow throat and breast, white abdomen and undertail-coverts.
Range Breeds in North America to central Mexico; winters mainly in Middle America to Panama. Rare on passage in the northern Bahamas and the only reports from the Greater Antilles are from Cuba (October, February and May), Hispaniola (October) and the Cayman Islands (February, April).
Status Vagrant or very rare passage migrant. Only recent record on Little Cayman, April 2006.

Clay-colored Sparrow   *Spizella pallida*
Taxonomy Monotypic.
Description L 12–13.5cm (4.5–5.25in). Adult has bright buffy-brown upperparts and rump, black streaks on back, grey nape and neck sides, whitish-buff underparts; face contrasts with white median crown-stripe, wide whitish supercilium and black malar stripe, pale lores; brown ear-coverts edged with dark brown, tail medium long and notched. Immature similar but with fine streaks on underparts.
Similar Species Chipping Sparrow has reddish-brown crown and grey-buff ear-coverts.
Range Breeds in northern North America and winters in southern Texas and Mexico. Vagrant to the Bahamas, Cuba and Cayman Islands, October–February.
Status Vagrant. One record on Grand Cayman, October 2011.

Savannah Sparrow   *Passerculus sandwichensis*
Taxonomy Polytypic (4).
Description L 13cm (5in). Adult has upperparts patterned brown and buff, pale crown stripe and wide supercilium, dark malar stripe edged with white, grey throat, underparts broadly streaked brown, bill large and thick with curved culmen, short notched tail, and pink legs.
Range Breeds and winters in North and Middle America; winters in the West Indies in the northern Bahamas, Cuba and Swan Islands.
Status Rare passage migrant. Five recent records in the Cayman Islands, flock of ten in Little Cayman in 1998 and single birds on Grand Cayman, October-November, 2001–2002. Previously an uncommon winter visitor and passage migrant, August-March in rough pasture and gardens.

White-crowned Sparrow   *Zonotrichia leucophrys*
Taxonomy Polytypic (7).
Description L 17cm (6.5in). Adult has crown boldly striped black and white, pointed pinkish or yellowish bill; grey nape, throat, breast and rump, brownish flanks and white undertail-coverts; dark streaking on mantle and two narrow white wing-bars on brown and buff wings. Immature has brown and buff crown-stripes.
Range Breeds in North America; *Z. l. leucophrys* winters in United States to northern Mexico and the West Indies: where it is rare in winter and on passage in Cuba and the Bahamas; vagrant to Jamaica.
Status Vagrant or rare passage migrant. Four records on Grand Cayman in fall, and one on Little Cayman in spring, 2000–2010.

Eastern Meadowlark   *Sturnella magna*
Taxonomy Polytypic (16).
Description L 23cm (9in). Heavy-bodied, short tail has white outer rectrices, long bill. Breeding adult has crown and upperparts streaked black and buff, wide pale supercilium, bright yellow throat and underparts with a broad black breast-band and darkly streaked flanks, pale legs. Non-breeding adult has brown streaking on crown and upperparts and less distinct breast-band.
Range Breeds in the United States, Middle America to South America, and in the West Indies in Cuba. Populations outside North America are sedentary; winters in the United States south throughout the breeding range; vagrant elsewhere.
Status Vagrant. One record on Cayman Brac, September 1987; no recent records.

Yellow-headed Blackbird   *Xanthocephalus xanthocephalus*
Taxonomy Monotypic.
Description L 25cm (10in). Adult male is black with orange-yellow head, throat and breast; black lores and white patch on wing-coverts. Adult female has brownish upperparts; yellow supercilium, moustachial stripe and breast, and dark eye-stripe. Immature has cinnamon-buff head, otherwise brownish with two pale wing-bars.
Range Breeds in North America; winters in south-western United States and northern Mexico; vagrant to the West Indies in the Bahamas, Cuba and Cayman Islands.
Status Vagrant. Five records on Grand Cayman, 1984 2009; an immature male was kept in captivity by a farmer for several months and released when in adult plumage. Observed in rough pasture.

SELECTED BIBLIOGRAPHY

American Ornithologists' Union. 1998. *Check-list of North American Birds*, 7th edition. American Ornithologists' Union. Allen Press, Kansas. With 54 supplements.

Bangs, O. 1916. A collection of birds from the Cayman Islands. *Bull. Mus. Comp. Zool.* 60(7): 303–320.

Bradley, P. E. 1986. A census of *Amazona leucocephala caymanensis*, Grand Cayman and *Amazona leucocephala hesterna*, Cayman Brac. *Cayman Islands Gov. Tech. Pub. No. 1*, George Town, Cayman Islands.

Bradley, P. E. 1994. The Avifauna of the Cayman Islands: an overview. Pages 377–407 *in* Brunt, M. A. & Davis, J. (Eds).*The Cayman Islands: Natural History and Biogeography.* Kluwer Acad. Publ., The Netherlands.

Bradley, P. E. 1995. *Birds of the Cayman Islands.* Photographs by Y-J. Rey-Millet. *2nd edition.* Caerulea Press, Italy.

Bradley, P. E. 2000. *The Birds of the Cayman Islands. B.O.U. Checklist No. 19.* British Ornithologists' Union, Tring, England.

Bradley, P. E., Cottam, M., Ebanks-Petrie, G. and Solomon, J. 2006. Cayman Islands Pp. 65–98 *in* Sanders, S. M. (Ed.). *Important Bird Areas in the UK Overseas Territories: priority sites for conservation.* Royal Society for the Protection of Birds, Sandy, Beds, U.K.

Bradley, P. E. 2009. The Cayman Islands. Pages 58–65 *in* Bradley, P. E. & Norton, R. L. (Eds). *An Inventory of Breeding Seabirds in the Caribbean.* University Press of Florida, Gainesville, Florida.

Burton, F. J. 2008. Vegetation Classification for the Cayman Islands. In *Threatened Plants of the Cayman Islands: the Red List.* Royal Botanic Gardens, Kew, U.K.

Burton, F. J., Bradley, P. E., Schreiber, E. A., Schenk, G. A. & Burton, R. E. 1999. Status of Red-footed Boobies *Sula sula* on Little Cayman, British West Indies. *Bird Conservation Inter.* 9: 227–233.

Clements, J. 2009. *The Clements Checklist of the Birds of the World.* www.birds.cornell.edu/clementschecklist

Cory, C. B. C. 1892. *The Birds of the West Indies.* Estes & Lauriat, Boston.

del Hoyo, J. A., Elliott, A. & Sargatal, J. (Eds). 1992–2002. *Handbook of the Birds of the World.* Vols. 1–7. Lynx Edicions. Barcelona.

del Hoyo, J. A., Elliott, A. & Christie, D. (Eds). 2003–2012. *Handbook of the Birds of the World.* Vols. 8–16. Lynx Edicions. Barcelona.

Diamond, A. W. 1980. The Red-footed Booby colony on Little Cayman: size, structure and significance. *Atoll Res. Bull.* 241: 165–170.

Garrido, O. H. & Kirkconnell, A. 2000. *Field Guide to the Birds of Cuba.* Cornell University Press, New York.

Hallett, B. 2006. *Birds of the Bahamas and the Turks and Caicos Islands.* Macmillan Caribbean, Oxford.

Haynes-Sutton, A., Downer, A., Sutton, R. & Rey-Millet, Y-J. 2009. *A Photographic Guide to the Birds of Jamaica.* Christopher Helm, London.

Johnston, D. W. 1975. Ecological analysis of the Cayman Islands avifauna. *Bull. Florida State Mus. Biol. Sci.* 19(5): 235–300.

Johnston, D. W., Blake, C. H. & Buden, D. W. 1971. Avifauna of the Cayman Islands. *Quart. J. Florida Acad. Sci.* 34 (2): 141–156.

Latta, S. C. Rimmer, C., Keith, A., Wiley, J. W., Raffaele, H., McFarland, K. & Fernandez, E. 2006. *Birds of the Dominican Republic and Haiti.* Princeton University Press, Princeton, N.J.

Lowe, P. R. 1911. On the birds of the Cayman Islands, West Indies. *Ibis* 53: 137–161.

Proctor, G. (in press). *Flora of the Cayman Islands.* 2nd Ed. Her Majesty's Stationery Office, London.

Raffaele, H. A., Wiley, J., Garrido, O., Keith, A. & Raffaele, J. A. 1998. *Guide to birds of the West Indies.* Princeton University Press, N. J.

Sibley, D. A., 2000. *The Sibley Guide to Bird*s. A.A. Knopf, New York.

Wiley, J. W., Gnam, R. S., Burton, F., Walsh, M., Weech, J., Strausberger, B. & Marsden, M. 1992. *Report on observations of the Cayman Brac Parrot* (Amazona leucocephala hesterna) *on Cayman Brac, June 1991.* Report for the International Council for Bird Preservation, Cambridge.

INDEX